A Clean Sheet Approach to Space Acquisition in ' the New Spac

WILLIAM SHELTON, CYNTHIA R. COOK, CHARLIE BARTON, FRANK CAMM, KELLY ELIZABETH EUSEBI, DIANA GEHLHAUS, MOON KIM, YOOL KIM, MEGAN MCKERNAN, SYDNE NEWBERRY, COLBY P. STEINER

Prepared for the Department of the Air Force
Approved for public release; distribution unlimited

For more information on this publication, visit **www.rand.org/t/RRA541-1**.

Published by the RAND Corporation, Santa Monica, Calif.

Library of Congress Cataloging-in-Publication Data is available for this publication.

ISBN: 978-1-9774-0744-3

Cover: Nmedia/Adobe Stock.

About This Report

Space is becoming increasingly competitive, with potential adversaries rapidly investing in new space capabilities with increasing technical expertise. At the same time, a revolution in space capabilities is being driven by both nontraditional suppliers—new commercial entrants to the space arena—and traditional defense contractors. The newest service, the U.S. Space Force (USSF), has a unique opportunity to take advantage of the widening spectrum of commercial capabilities and create new management processes to respond to the challenge presented by potential adversaries in space. To support this effort, Space and Missile Systems Center (SMC) leadership asked RAND Project AIR FORCE to develop a "clean sheet" acquisition approach designed around the new service's unique mission and calling. Our recommendations derive from our analysis of the available literature combined with interviews with more than 45 current and retired senior leaders and space acquisition experts, most with several decades of acquisition and/or operations experience. This report should be of interest to USSF and broader Department of the Air Force (DAF) leadership and those involved in weapon system acquisition across the U.S. Department of Defense.

The research reported here was commissioned by SMC and conducted within the Resource Management Program of RAND Project AIR FORCE as part of a fiscal year 2020 project titled *Updating Space Acquisition in Light of the New Space Force*. Research was conducted from January 2020 through December 2020.

RAND Project AIR FORCE

RAND Project AIR FORCE (PAF), a division of the RAND Corporation, is the Department of the Air Force's (DAF's) federally funded research and development center for studies and analyses, supporting both the United States Air Force and the United States Space Force. PAF provides the DAF with independent analyses of policy alternatives affecting the development, employment, combat readiness, and support of current and future air, space, and cyber forces. Research is conducted in four programs: Strategy and Doctrine; Force Modernization and Employment; Manpower, Personnel, and Training; and Resource Management. The research reported here was prepared under contract FA7014-16-D-1000.

Additional information about PAF is available on our website:

www.rand.org/paf/

This report documents work that was originally shared with the DAF on October 19, 2020. The draft report, issued on January 8, 2021, was reviewed by formal peer reviewers and DAF subject-matter experts.

Contents

Figures

Summary

Issue

Space is becoming increasingly competitive, with potential adversaries rapidly investing in new space capabilities with increasing technical expertise. At the same time, a revolution in space capabilities is being driven by both nontraditional suppliers—new commercial entrants to the space arena—and traditional defense contractors. As the U.S. Space Force (USSF) is stood up in response to the threat, the new service has a unique opportunity to take advantage of the widening spectrum of commercial capabilities and create new management processes to respond to the challenge. To support this effort, Department of the Air Force (DAF) leadership asked RAND Project AIR FORCE to develop a "clean sheet" acquisition approach designed around the new service's unique mission and calling.

Approach

We intentionally focused our research on internal USSF acquisition, acknowledging that there are multitudinous external interfaces—e.g., the other services, the Missile Defense Agency, and the Intelligence Community (IC). These recommendations are derived from our analysis of the literature combined with interviews with more than 45 current and retired senior leaders and space acquisition experts, most with several decades of acquisition and/or operations experience.

Findings

The new acquisition vision links to several USSF features that are either unique—setting it apart from the other services—or particularly pronounced. Most notably, USSF will be significantly smaller than any other military service, by more than an order of magnitude. This means that there will be fewer people for processes that traditionally have been manpower intensive, including acquisition. A smaller service offers the opportunity for increased agility and the reduced bureaucracy resulting from a flatter organization and a shorter chain of command. Second, USSF is highly reliant on technology to develop and sustain its joint warfighting capabilities, perhaps even more so than other services. USSF warfighters are technology operators and have much in common with space acquirers. This dependence on technology necessitates USSF having a close, trusting, collaborative relationship with industry.

A service having technology as a foundation for warfighting warrants an acquisition approach focused on ensuring that the required capabilities are available when needed. To be effective in this context, acquisition processes must be rapid, agile, and, above all, threat informed. We offer a new "clean sheet" acquisition vision[1] for the technology-centric USSF—*acquisition as a warfighting capability* rather than a *support function*. The additional benefit to focusing on the execution of an effective capability is that, although potential adversaries can exfiltrate weapon systems technology, effectively copying an approach based on a strong culture is a much bigger challenge. Also, a new space acquisition culture enabled by increased agility and a shorter chain of command, integrated with space warfighting, will enable threat-focused, innovative space capabilities to maintain and strengthen the United States' advantage in space.

Recommendations

To provide threat-informed capability on an operationally viable schedule within cost constraints, the clean sheet approach of acquisition as a warfighting capability incorporates these features:

- **Remove the seams traditionally separating operators and acquirers** so all understand both technology and operations; operators will know how technology flows and changes, and acquirers will know how the technology is implemented and used in the field.
- **Create an adaptive technical architecture**, based on warfighting doctrine and concept of operations, to provide a framework for decisionmaking, countering the threat, and a road map for innovation.
- **Establish a single space acquisition decisionmaker** for flexible management of the enterprise—focusing resources on the highest priorities, driving capability synchronization, and radically delegating to empowered subordinates.
- **Ensure a workforce consisting of experts** cultivated through selective recruiting, assignments, training, and promotions to be risk tolerant, flexible, collaborative, and enterprise-focused–providing capabilities, not merely systems.
- **Build internal and external outreach mechanisms**, including information-sharing and metrics, that emphasize strong relationships and mutual trust within and across the U.S. Congress, the U.S. Department of Defense, USSF, the IC, other federal agencies, and industry.
- **Foster a trusting, collaborative relationship with industry**—for example, by providing industry with a technology road map that includes (1) *innovation on-ramps* to accept emerging technology or address changing threats and (2) *divestiture off-ramps* for obsolete capability.

[1] This clean sheet vision for acquisition embodies a systematic and comprehensive approach rather than providing a menu of items from which to pick and choose.

This clean sheet vision for acquisition embodies a systematic, comprehensive, and holistic approach rather than providing a menu of items from which to pick and choose. USSF needs the flexibility and authority to invest in *all* of these changes across the enterprise—and Congress will need to provide the required authorities, including enhanced funding flexibility to allow for investments and disinvestments as the architecture evolves. All of this can and should be done holistically and intentionally to create the right culture and ensure effective change. The first Chief of Space Operations offered a vision for change at the 2020 Air Force Association Air, Space and Cyber Conference: "If we get this right, we will be the envy of the other services, because we are not tied to business of the past."[2]

[2] Charles Pope, "Driven by 'A Tectonic Shift in Warfare' Raymond Describes Space Force's Achievements and Future," Air Force Public Affairs, September 15, 2020.

Acknowledgments

The RAND Project AIR FORCE (PAF) project team would like to thank Lt Gen John "JT" Thompson, U.S. Air Force, for his sponsorship and insightful guidance throughout. Col Timothy Sejba, Lt Col Andy Anderson, and Noble Smith provided invaluable insight and access to information and U.S. Space Force (USSF) thinking. We are particularly grateful to the more than 45 senior leaders inside and outside USSF who spoke with us on a nonattribution basis, generously giving their time and their insights for this effort. Inputs from our peer reviewers, Bonnie Triezenberg (RAND), John Ausink (RAND) and Lara Schmidt (Aerospace), strengthened the report. Finally, we thank Obaid Younossi; Anu Narayanan; Patrick Mills, associate director of the Resource Management Program; and Ted Harshberger, former vice president of PAF.

CHAPTER ONE

Background

Policymakers have increasingly recognized that space is no longer an uncontested domain and instead now is an environment where potential adversaries are becoming more active and capable. To focus efforts on defending U.S. interests in space, the fiscal year (FY) 2020 National Defense Authorization Act (NDAA) called for the formation of a space force, separate from the U.S. Air Force, which had focused on the use of space to provide secure communication, weather, and precision timing and navigation services for the terrestrial fight.[1] The U.S. Space Force (USSF) was stood up as a separate service on December 20, 2019, with the signing of the USSF Act, as part of the 2020 NDAA.[2]

The establishment of a new service is an extremely rare event, with the U.S. Air Force being the last service stood up (in 1947). The challenges of managing a complex service are known, but ingrained culture and processes and existing interests make change difficult and slow. However, starting a new service offers a unique opportunity for creating a different culture and incorporating innovative ideas and management structures.

One notable challenge in the space domain has been the acquisition of space capabilities, including satellites and their components, user equipment, ground stations, and launch services. Many space systems, especially satellites, can be described as exquisite, monolithic, highly technical solutions. These solutions often were the result of specific operational capability needs necessitating extremely high accuracy and reliability.[3] Technical challenges, schedule slippage, and cost growth create challenges in delivering necessary capabilities on time and on budget. As weapon systems across the entire U.S. Department of Defense (DoD) become increasingly interconnected (and, in many cases, reliant on space capabilities), not

---

[1] Public Law 116-92, National Defense Authorization for Fiscal Year 2020, December 20, 2019.

[2] USSF, "About Space Force," webpage, undated.

[3] One such example is the Advanced Extremely High Frequency (AEHF) satellite, a joint service satellite communications (SATCOM) system providing global, survivable, secure, protected, and jam-resistant communications for high-priority military ground, sea, and air assets. AEHF satellite allows the National Security Council and Combatant Commanders to control their tactical and strategic forces at all levels of conflict, up to and including general nuclear war, and it supports the attainment of information superiority. Another is the Space-Based Infrared System (SBIRS) High (SBIRS High). SBIRS High is an integrated system consisting of multiple space and ground elements with incremental deployment phasing, simultaneously satisfying requirements in the following mission areas: missile warning, missile defense, technical intelligence, and battlespace awareness.

having these weapons available to the warfighter when needed could put DoD's ability to achieve the desired strategic outcomes at risk.

To address these long-standing challenges, the Department of the Air Force (DAF) asked RAND Project AIR FORCE to develop an unconstrained "clean sheet" approach to space acquisition. That effort is described in this report. Our guiding principles for the effort were the Chief of Space Operation's (CSO's) formal guidance and public statements that USSF acquisition should be informed by the threat and should be innovative.[4] Other components of the vision include responsiveness—speed and agility. Finally, consciousness of resource limitations is necessary, given that defense budgets are limited. Transformational change is the goal of the CSO—Gen John Raymond—for improving USSF acquisition: "If we get this right, we will be the envy of the other services because we are not tied to the past."[5]

What Is Different About the Needs of the Space Force?

The DAF space-focused acquisition system has encountered numerous challenges in conceptualizing, developing, fielding, and maintaining vital space systems on time and budget. Space systems are heavily technology driven, and this technology is evolving at an increasingly fast pace. This primacy of technology drives the need for acquisition teams and operators with advanced technical training and experience. Furthermore, potential adversaries are increasing their investments and working to develop advanced space capabilities. Thus, the pace of technological progression that is needed to maintain any advantage over potential adversaries in space increases the need for DoD to draw on the commercial space industry, particularly nontraditional suppliers, such as small startups, which lead the way in technological innovation.

More than other warfighting domains, domination of space necessitates highly orchestrated horizontal and vertical synchronization. Horizontal synchronization is necessary to enable interaction among legacy and newer systems, cross-service participation in space efforts, and critical involvement of other agencies (e.g., the National Reconnaissance Office [NRO] and the Missile Defense Agency). Vertical synchronization is critical for ground and space systems interoperability, an area that has suffered inadequate attention in the past, as we discuss later in this chapter. (Note that *synchronization* is the vertical and horizontal alignment of space assets across the space enterprise.[6] *Vertical synchronization* is integration

[4] John Raymond, *Chief of Space Operations' Planning Guidance*, Washington, D.C.: U.S. Space Force, November 9, 2020.

[5] Charles Pope, "Driven by 'A Tectonic Shift in Warfare' Raymond Describes Space Force's Achievements and Future," Air Force Public Affairs, September 15, 2020.

[6] We use the term *space enterprise* from an acquisition point of view, and it encompasses all the systems that are brought together to provide capabilities for the warfighter.

and alignment of segments within a space mission. *Horizontal synchronization* is integration and alignment across space missions.)

All of these challenges demand a new approach to acquisition that is fast, agile, and tolerant to risk[7] or even failure. Acquisition should not be considered as playing a supporting rather than a warfighting role, and acquisition professionals must work seamlessly with operators to ensure that they have the needed capabilities reliably and on time.

Challenges of Past and Current Space Acquisition

DoD space acquisition has a long history of difficulties related to technical challenges, high costs, and long development timelines. Some space programs, such as SBIRS, AEHF satellite, global positioning system (GPS) IIF, and GPS III, have taken ten years or longer from Milestone B to the first satellite launch of their constellations, and all of them have experienced major cost growth.[8] Ground systems within the space enterprise do not fare much better. The Next Generation Operational Control System (OCX), the modernized ground control system for GPS, continues to struggle to deliver the capabilities needed to use GPS III's modernized signals. The Joint Space Operations Center Mission System (JMS), a space situational awareness and space command and control system, also experienced many technical difficulties and schedule slips before it was eventually canceled almost ten years after the program's start.[9]

Such delays lead to synchronization issues within the space enterprise, leaving many modernized capabilities untapped for years until the associated control system or the user equipment are fielded. In addition to the OCX delays, the development of GPS user equipment receiver cards for the modernized military signals has experienced multiple delays.[10]

[7] Risk must be addressed situationally. In an operational context, there is the term *acceptable risk*. In an acquisition context, this would mean that more risk would be acceptable when responding to an urgent, emerging requirement than during a traditional acquisition. However, acquirers must understand that avoiding all risk in an acquisition and the resulting delays could have more-dire outcomes than taking calculated risks and potential failures. Our use of *risk tolerance* is intended to define a state of mind in which acquirers lean forward and failures are not only acceptable but also understood as learning opportunities.

[8] Yool Kim, Elliot Axelband, Abby Doll, Mel Eisman, Myron Hura, Edward G. Keating, Martin C. Libicki, Bradley Martin, Michael McMahon, Jerry M. Sollinger, Erin York, Mark V. Arena, Irv Blickstein, and William Shelton, *Acquisition of Space Systems, Volume 7: Past Problems and Future Challenges*, Santa Monica, Calif.: RAND Corporation, MG-1171/7-OSD, 2015; DoD, "Selected Acquisition Report: Global Positioning System III (GPS III)," December 2018.

[9] U.S. Government Accountability (GAO), *Space Command and Control: Comprehensive Planning and Oversight Could Help DOD Acquire Critical Capabilities and Address Challenges*, GAO-20-146, October 2019d.

[10] GAO, "Space Acquisitions: DOD Faces Significant Challenges as It Seeks to Address Threats and Accelerate Space Programs," statement of Cristina T. Chaplain, director of Contracting and National Security Acquisitions, before the Subcommittee on Strategic Forces, Committee on Armed Services,

Similarly, improved space surveillance capabilities offered by Space Fence, deployed in 2018, cannot be fully leveraged until a modernized space situational awareness data-processing system comes online, which JMS was supposed to deliver.[11]

These challenges and shortcomings in space acquisitions have been attributed to a set of broader issues related to leadership, organizational management, and culture, according to multiple GAO reports, congressionally directed studies, past RAND reports, academic articles,[12] and DoD officials. One persistent issue is the fragmentation of acquisition authorities in the space enterprise. The 2008 Allard commission found that the "leadership for strategy, budgets, requirements, and acquisition across NSS [national security space] is fragmented, resulting in an absence of clear accountability and authority."[13] A 2016 GAO report found eight organizations that have space acquisition management responsibilities across services and the NRO, 11 organizations that have space oversight responsibilities, and six that are responsible for establishing requirements for space programs.[14] A related issue has been the lack of unified strategy, plans, and architecture for the space enterprise, which help guide decisions within the space acquisition management processes (requirements, budgeting, and acquisition). With multiple organizations having roles and responsibilities in those areas, coordinating and integrating across numerous seams has proven to be lengthy and challenging.[15]

In the past, the significant investment to put a capability in space had often driven space acquisition to focus on the longevity and exquisite[16] performance of individual systems, which increased costs and fostered a failure-intolerant culture, which was also focused on individual systems or programs instead of enterprise capabilities.[17] However, recognizing that the current space enterprise is not sufficiently resilient in the face of growing space threats, the Space Enterprise Vision in 2016 highlighted that "acquisition and programmatic decisions

House of Representatives, GAO-19-482T, April 3, 2019b; GAO, "DOD Space Acquisitions Management and Oversight: Information Presented to Congressional Committees," GAO-16-592R, July 27, 2016b.

[11] GAO, *Weapon Systems Annual Assessment: Limited Use of Knowledge-Based Practices Continues to Undercut DOD's Investments*, GAO-19-336SP, May 2019c.

[12] Ellen Pawlikowski, Doug Loverro, and Tom Cristler, "Space: Disruptive Challenges, New Opportunities, and New Strategies," *Strategic Studies Quarterly*, Vol. 6, No. 1, Spring 2012.

[13] A. Thomas Young, Edward Anderson, Lyle Bien, Ronald R. Fogleman, Keith Hall, Lester Lyles, and Hans Mark, *Leadership, Management, and Organization for National Security Space: Report to Congress of the Independent Assessment Panel on the Organization and Management of National Security Space*, Alexandria, Va.: Institute for Defense Analyses, 2008.

[14] GAO, 2016b.

[15] GAO, 2016b.

[16] *Exquisite* is defined as "large, highly sophisticated satellites usually based in GEO [geosynchronous orbit] for imagery, missile warning and Intelligence Community missions" (Theresa Hitchens, "Wilson: DoD Study Finds 'Exquisite' Satellites Still Needed," *Breaking Defense*, April 9, 2019).

[17] Kim et al., 2015; DoD, 2018.

can no longer occur in mission area stovepipes, but must instead be driven by an overarching space mission enterprise context."[18] Several initiatives ensued aimed at creating a resilient space enterprise that would include acquisition reform efforts. The Space and Missile Systems Center (SMC) underwent its largest organizational restructuring to enable management of space acquisition as an enterprise.[19] As USSF aims to create an organization that can outpace space threats and transform space acquisition to support that goal, it must tackle these leadership, organizational management, and cultural issues head-on.

Research Approach

Guidance from the Space Force

When our sponsors asked us to develop a clean sheet approach to space acquisition, they directed us to focus on several specific topics to guide our research, termed *sprints*. The initial five sprint topics were (1) an overall vision for the clean sheet acquisition, (2) acquisition governance, (3) capability synchronization, (4) workforce development and management (which we term *talent management*), and (5) interfacing with industry. On mutual agreement, we added a sixth sprint on change management to ensure that implementation issues were deliberately considered. Given the extent of space acquisition and our available resources, we intentionally focused our research on internal USSF acquisition, acknowledging that there are multitudinous external interfaces (e.g., the other services, the Missile Defense Agency, and the Intelligence Community (IC).

Literature Review

Our literature review included senior-level statements and writings; acquisition policy and guidance; and information on space programs from a variety of sources, including official government documentation, the GAO, published RAND reports, other federally funded research and development centers (FFRDCs), trade literature, and literature on commercial best practices and change management.

Development of Framing Assumptions

Using the literature and team subject-matter expertise, each sprint team identified foundational assumptions that, if not true, would cause the program to fail—also known as "framing assumptions."[20] Acquisition guidance defines *framing assumptions* as "any supposition

---

[18] Air Force Space Command Public Affairs, "Hyten Announces Space Enterprise Vision," April 13, 2016.

[19] Sandra Erwin, "SMC 2.0: Air Force Begins Major Reorganization of Acquisition Offices," *SpaceNews*, April 17, 2018.

[20] Mark Husband, "Information Paper on Framing Assumptions," U.S. Department of Defense, September 13, 2013; and Mark V. Arena, Irv Blickstein, Abby Doll, Jeffrey A. Drezner, James G. Kallimani, Jennifer Kavanagh, Daniel F. McCaffrey, Megan McKernan, Charles Nemfakos, Rena Rudavsky, Jerry M. Sollinger, Daniel Tremblay, and Carolyn Wong, *Management Perspectives Pertaining to Root Cause Analyses*

(explicit or implicit) that is central in shaping cost, schedule, or performance expectations of an acquisition program."[21] Key features are that they significantly affect program expectations, have consequences that cannot be easily mitigated, are not derivative of other assumptions, and do not apply generically to all programs. If framing assumptions are proven false, they will prevent successfully developing and implementing potential approaches for improving space acquisition.[22] Our efforts to define framing assumptions started by identifying desired characteristics of clean sheet space acquisition, which are that it is (1) threat-informed, (2) operationally responsive, and (3) cost conscious. These characteristics informed the specific framing assumptions, developed by each sprint and described later in this report, and guided the development of interview questions.

Discussions and Interviews with Subject-Matter Experts

Our team held more than 45 separate discussions with subject-matter experts (SMEs) with a wide variety of relevant backgrounds, conducted on a nonattribution basis. Participants included active and retired senior military leaders, including numerous general officers; current and retired senior civilians, including several in the Senior Executive Service; experts with industry experience; and individuals from FFRDCs. Figure 1.1 indicates the organizational background of the participants.

Because of the breadth of topics covered in this study and the varied backgrounds of the participants, we did not use a single protocol for all the discussions. Each discussion began with an overview of the project and introductions, followed by an open question in which we elicited ideas for improving space acquisition. Depending on the background of the interviewee, we asked questions relating to governance, synchronization of space assets, talent management, industry interaction, or change management. Although we looked for any broad trends across the answers, we also looked for promising and implementable ideas, even if offered by only one or two participants.

Caveats: Response to COVID-19

Remote working combined with limitations on travel meant that our team focused on unclassified data collection that could be conducted using open-source materials and on unclassified telephone calls with a distributed RAND team participating from multiple locations. Additionally, all interviews were conducted virtually.

---

of Nunn-McCurdy Breaches, Volume 4: Program Manager Tenure, Oversight of Acquisition Category II Programs, and Framing Assumptions, Santa Monica, Calif.: RAND Corporation, MG-1171/4-OSD, 2013.

[21] Husband, 2013.

[22] Framing assumptions are inherently the identification of sources of risk because, if the framing assumptions are false, then the desired outcomes will not occur. Our project scope did not allow us to conduct a full-scale assessment of the likelihood of our framing assumptions being supported or the impact on our clean sheet approach if they were not true. Future research could conduct such a review and identify areas of risk and related mitigations.

FIGURE 1.1
Organizational Background of Participants

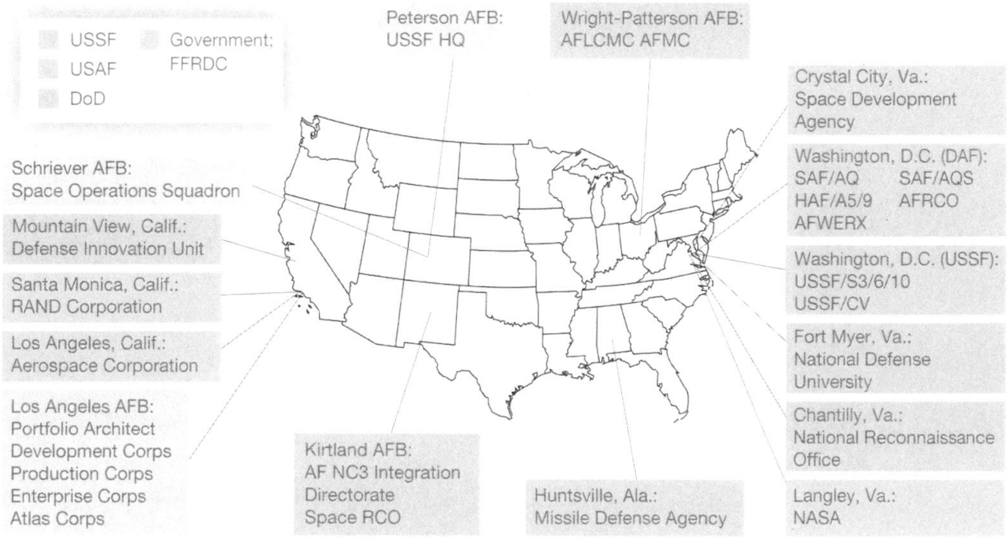

USSF
USAF
DoD
Government: FFRDC

Peterson AFB: USSF HQ
Wright-Patterson AFB: AFLCMC AFMC
Crystal City, Va.: Space Development Agency
Washington, D.C. (DAF): SAF/AQ SAF/AQS HAF/A5/9 AFRCO AFWERX
Washington, D.C. (USSF): USSF/S3/6/10 USSF/CV
Fort Myer, Va.: National Defense University
Chantilly, Va.: National Reconnaissance Office
Langley, Va.: NASA
Huntsville, Ala.: Missile Defense Agency
Kirtland AFB: AF NC3 Integration Directorate Space RCO
Schriever AFB: Space Operations Squadron
Mountain View, Calif.: Defense Innovation Unit
Santa Monica, Calif.: RAND Corporation
Los Angeles, Calif.: Aerospace Corporation
Los Angeles AFB: Portfolio Architect Development Corps Production Corps Enterprise Corps Atlas Corps

NOTE: AFB = air force base; HQ = headquarters; USAF = U.S. Air Force.

Structure of This Report

Following this introductory chapter, Chapter Two proposes a vision for a strategic approach to USSF acquisition. Chapter Three presents a vision of acquisition decisionmaking to improve governance and capability synchronization. Chapter Four discusses talent management. Chapter Five offers ideas on how to effectively engage with industry. Chapter Six contains findings, recommendations, and an approach to change management and implementation. Chapter Seven offers summary conclusions and recommendations. A list of references rounds out the report.

CHAPTER TWO

The Clean Sheet Vision: Acquisition as a Warfighting Capability

Introducing the Vision

Over the past few years, policymakers have increasingly emphasized that space is no longer an uncontested sanctuary, free of potential military conflict. In his initial planning guidance, the first CSO, Gen John "Jay" Raymond, summarized the challenge:

> The convergence of proliferating technology and competitive interests has forever redefined space from a benign domain to one in which we anticipate all aspects of human endeavor—including warfare. The return of peer, great power competitors has dramatically changed the global security environment and space is central to that change.[1]

Similarly, the 2018 National Defense Strategy[2] indicates that "new threats to commercial and military uses of space are emerging" (p. 3) and describes space as a warfighting domain (p. 6). The investments of near-peer competitors—potential adversaries—combined with the information-centric nature of space operations mean that operations can be ongoing in the absence of a shooting war. The increasing availability of capabilities (such as communications and imaging) developed and offered by nontraditional commercial companies and the rapidly decreasing commercial launch costs have reduced barriers to entry for any entity with the wherewithal to purchase the desired services or items. In the absence of extended active operations, this may be a return to an era of "boiling peace."[3]

Although all modern warfare takes advantage of technology to some extent, space warfare might do so even more. Space warfighting doctrine will necessarily focus on the implementation of technology—either better or more effectively than potential adversaries. As guardians develop their doctrine and plan for the future, they will benefit from understanding where

---

[1] Raymond, 2020, p. 3.

[2] Jim Mattis, *Summary of the 2018 National Defense Strategy of the United States of America: Sharpening the American Military's Competitive Edge*, Washington, D.C.: U.S. Department of Defense, 2018.

[3] Jeanne D. Heller, ed., *Project AIR FORCE 1999 Annual Report*, Santa Monica, Calif.: RAND Corporation, AR-7042-AF, 2000.

technology is moving, both in the commercial world and in investments by potential adversaries.[4] Doctrine development and operational planning will be more effective if space warfighters understand technology trends. Because the necessity for this derives from increasing the capabilities of potential adversaries, these activities must be "threat-informed," requiring connections with the IC.

Traditionally, warfighters have set requirements based on what they need—identified capability gaps—and DoD used these to drive innovation. In recent policy statements, senior leaders have suggested that traditional acquisition models have led to large exquisite programs: "When the space age began, mastery of the most advanced disciplines of science, engineering, and manufacturing was required to produce a few exquisite systems."[5] Contractors developed these systems in response to requirements set by their government customers. But just as the strategic environment has changed, with the advent of new competitors in space, the commercial environment has also changed. The Chief of Staff of the Air Force, General Charles Q. Brown, Jr., recently recognized this factor: "Unlike the past, much of the emerging technologies that will determine our future are no longer created or funded by the DoD."[6] This represents a more concerning challenge, as then–Secretary of Defense James Mattis affirmed:

> The fact that many technological developments will come from the commercial sector means that state competitors and non-state actors will also have access to them, a fact that risks eroding the conventional overmatch to which our Nation has grown accustomed.[7]

That technological developments are coming independently from the commercial sector rather than as a result of requirements-setting means that DoD is not the sole, or potentially the primary, driver of technological change. Nontraditional suppliers are breaking ground in space launch vehicles and services, satellites, satellite constellations, and information systems. This offers both an opportunity and a challenge. The opportunity is in the adoption of technological innovations from commercial firms,[8] especially the nontraditional suppliers that may be new entrants to the space domain. The challenge is that operators will need to have insight into and a technical understanding of how technology is changing to set require-

[4] Understanding the capabilities of potential adversaries requires input from and close coordination with the IC.

[5] Raymond, 2020, p. 3.

[6] Charles Q. Brown, Jr., *Accelerate Change or Lose*, Washington, D.C.: U.S. Air Force, August 31, 2020.

[7] Mattis, 2018, p. 3.

[8] There are several different ways USSF could adopt commercial technologies, such as purchasing commercial services or using commercial off-the-shelf services. Determining how USSF should leverage or integrate innovative commercial technologies into its architectures or acquisition programs depends on a multitude of factors, and it is an ongoing area of analysis in USSF.

ments that take advantage of this innovation.[9] A traditional model of identifying warfighting gaps and generating requirements from those gaps without understanding technology trends and flows will no longer yield the greatest value—or the most useful technology—from the investments. Similarly, warfighters developing doctrine and plans for future space engagements need an understanding of commercial trends to ensure that those plans make the best use of the capabilities that will be available. If warfighters are separated from acquisition—the community[10] that most clearly sees and understands commercial capabilities—they are at a disadvantage when it comes to developing and updating doctrine, plans, and requirements.

Likewise, the traditional role of acquisition professionals—offering "support" to warfighters from across an organizational and functional seam—will also no longer yield the best results. Acquisition cannot be divorced and separate from space operations: Acquisition professionals need regular, if not continuous, interaction with warfighters to understand trends in doctrine and plans and to identify gaps, so that they can more effectively perform their market research and articulate how commercial offerings will fill these gaps. Although this is not a new idea—operators and intelligence personnel have always informed acquisition—this input has come from outside the acquisition community, across the seam. Just as requirements cannot be divorced from understanding technology trends, acquisition needs to be deeply imbued with operational knowledge and culture. Getting capabilities at the speed of relevance will be necessary for effectiveness in meeting the challenges posed by our potential adversaries.

Our review of senior leader guidance, combined with our interviews, supports the idea that a dramatically new approach to space acquisition is necessary, or the challenges summarized in Chapter One are likely to persist, and even be exacerbated. Acquisition is typically viewed as a support function—as part of the "tail" that supports the "tooth" (i.e., the warfighter). Furthermore, acquisition has often demonstrated that it has a risk-averse culture, not a risk-tolerant one, thereby deepening the seam between it and the operational community. The need for change goes beyond acquisition reform, which, while important, may not be sufficient: Reform has often focused on fixing acquisition, not re-creating it.

To effect real change, acquisition needs to be reimagined and understood as a *warfighting capability* critical to the mission. This view needs to become a part of USSF's culture—and not solely within the traditional acquisition community, but also within space operations. Acquisition professionals and space operators are taking on different aspects of space technology management. Notably, this idea was supported by the majority of our interviewees.

This culture change necessary to reframe the role of acquisition will be significant, but it can be instantiated internally by removing the seam that currently separates the opera-

---

[9] Although the overall requirements process may also benefit from a clean sheet review, the full variety of activities covered in requirements setting was not part of this research.

[10] This view of the acquisition community covers such organizations as defense laboratories, the Defense Innovation Unit, AFWERX, and the Defense Advanced Research Projects Agency, which feed into acquisition through the technology transition process.

tion and the acquisition communities, a concept that will be discussed in greater detail in Chapter Two. This concept received broad support in our interviews, and not just from the acquisition community, as we expected, but also from the space operations community. Externally, industry engagement will also need to be reimagined to better connect partners and solutions with operators and their needs.

Removing the seam has organizational management implications for USSF. In July 2020, USSF organizational structure was announced to consist of three field commands. Space Systems Command (SSC) will be responsible for all aspects of acquisition. Space Operations Command (SpOC) will be responsible for warfighting (more specifically, to provide warfighting forces and capabilities to combatant commands and others during active operations). Space Training and Readiness Command (STARCOM) will be responsible for talent management. Operators and acquirers will need to be integrated across the organization structure. One approach is to rethink the naming of the first two of those three commands. SpOC's focus on posturing for warfighting could center on current operations, so notionally it could be the "Current Operations Command." SSC's focus on ensuring the availability of innovative capabilities for the future suggests a "Future Operations Command" framework. This is an imperfect framework, in part because SSC's sustainment and maintenance mission is near-term-focused—but it does highlight the operational importance of SSC's contribution to space warfighting. Although changing the commands' names might not initially seem like it would have a significant impact, as USSF develops and sustains its culture, these titles would reinforce the new approach.

We close this discussion of our clean sheet vision by noting that acquisition as a warfighting capability approach is not only implicit in recent senior guidance, as described earlier in this report, but it has been recognized as important in the past. In 1964, the historian Irving Brinton Holley, Jr., offered, "buying Aircraft presents one thesis above all others: the procurement process itself is a weapon of war no less significant than the guns, the airplanes, and the rockets turned out by the arsenals of democracy."[11] One of our interviewees noted that although specific technologies can be stolen or exfiltrated, an effective culture is much harder to imitate. Investing in and developing culture and processes specific to USSF—and space acquisition in particular—has the potential to be one of the most-effective tools for countering potential adversaries. More broadly, the interviews revealed strong support for reconsidering acquisition's role and removing the seam between the communities, with some interviewees expressing the need to rethink the requirements process and its connection to acquisition.

[11] Irving Brinton Holley, Jr., "Some Concluding Observations on Military Procurement," in *United States Army in World War II, Buying Aircraft: Matériel Procurement for the Army Air Forces*, Washington, D.C.: Office of the Chief of Military History, Department of the Army, 1964, p. 569.

Clean Sheet Framing Assumptions

The framing assumptions for the overarching clean sheet vision focus on three fundamental issues. The first is that legislative language would support an effective USSF and allow a transition to our clean sheet approach. In this effort we did not undertake a complete review of existing policy to identify specific barriers, in part because the final approach has not been decided by USSF, but also because there can be flexibility in the application in legislation that is clear to direct practitioners. Barriers may also evolve over time. What will be required is the willingness of USSF leaders to identify the needs and press for change. Some of what is needed is identified later in this report—but there may be more required.

The second is that the space architecture is not fixed and can be evolved over time to support a clean sheet approach. This may involve moving away from large "exquisite" systems to more modular and fault-tolerant constellations or other approaches that can be upgraded more regularly, to include low earth orbit (LEO) "megaconstellations" as are emerging in industry. This will require dedication on the part of USSF to ensure that requirements development aligns with the clean sheet vision.

Our final clean sheet assumption is that although USSF does not control either the threat or the pace of technical development, it can change its own processes to effectively outpace the threat and engage with industry to take advantage of new developments. The recommendations in this report are intended to support USSF's ability to meet those challenges.

Developing the Vision

To create and effectively transition to a new acquisition approach and culture, USSF needs to be as independent from the U.S. Air Force as possible. Heritage DAF and U.S. Air Force processes have created the culture that shapes and is appropriate for Air Force acquisition. Organizational change is notably difficult, so the creation of a separate service offers the opportunity to start off with something new. USSF should be allowed broad leeway and discretion in setting up acquisition policies and processes that best serve the space domain, regardless of how these deviate from the U.S. Air Force's policies and processes.

Decisionmaking independence is key, and in the DoD, as in many organizations, control over resources is the foundation of real authority. USSF will need budget independence and flexibility to be able to independently set its priorities, make investments, and allocate resources to its most important priorities. Enhanced budget flexibility will likely require new approaches to data transparency with stakeholders, especially Congress. Decisionmaking in a smaller, more agile service will be facilitated by a flatter hierarchy and reduced decision reviews. These approaches should allow for improved vertical and horizontal synchronization of capabilities. Finally, capability investments and disinvestments must align with an overarching space architecture ultimately managed by a single space acquisition decisionmaker. This will allow focusing of resources to achieve the best value for the enterprise and drive capability synchronization while radically delegating to empowered subordinates. Management and decisionmaking will be discussed in greater detail in Chapter Three.

Ensuring that acquirers and operators all understand how to use space technology and what the commercial world offers requires a USSF-specific approach to personnel development that also must be independent from, or at least freely able to deviate from, the U.S. Air Force. An approach that received significant support from the majority of our interviewees (including individuals from both the space operations and space acquisition communities) was to bring in all new entrants to USSF and give them assignments rotating between SSC and SpOC. Uniformed operators and acquirers would be in a common group for the first years of their career—perhaps until promotion to field grade officers, with comparable career paths for enlisted personnel.[12] For officers, the goal would be to recruit and develop what one interviewee termed "warrior engineers"[13] to ensure that they had the skill sets allowing them to understand technology both from an operator's and an engineer's or acquirer's perspective. Details of talent management are described in more detail in Chapter Four.

USSF is dependent on commercial industry to develop and produce space-related products and services. USSF personnel will need to be aware of technology developments and capabilities and how these fit together in complex systems, kinetic warfare, and effective two-way communication with industry and also how to interface with agility and focus on technology transition to onboard new capabilities. The challenge will be to maintain focus on how individual systems contribute to enterprise capabilities that align with an overarching space architecture. Chapter Five describes specifics.

Finally, we note that the clean sheet vision will not happen without leadership support. As Chapter Six describes, USSF must engage in thoughtful change management to evolve the U.S. Air Force approaches from heritage to the new clean sheet vision.

Summary

The descriptions in this chapter and the report's later chapters describe recommendations that together would yield a new clean sheet approach to acquisition as a warfighting capability. It is a systematic, comprehensive, and holistic approach rather than providing a menu of items from which to pick and choose. USSF needs the flexibility and authority to invest in *all* of these changes across the enterprise—and Congress will need to ensure that USSF has and maintains the required authorities, including enhanced funding flexibility, to allow for investments and disinvestments as the architecture adapts and evolves.

---

[12] The specific timing of these assignments requires additional research beyond the scope of this project.

[13] This term resonated with the project team and other interviewees, so we adopted it to cover individuals with the operator/acquirer perspective.

CHAPTER THREE

Rapid, Enterprise-Centric Decisionmaking

Rapid delivery of space capabilities will be critical to ensuring operational effectiveness. In this chapter, we describe how enhancements in governance and synchronization can improve space acquisition enterprise outcomes.

Rapid Decisionmaking Challenges

The United States' space enterprise has been ahead of its potential adversaries for decades in space capabilities. This advantage and lack of competition have allowed the space community to prioritize performance over speed. Consequently, as described in Chapter One, the space community has settled into a set of comfortable habits over the past 20 years that need to be changed for the United States to stay ahead of its potential adversaries in this now hotly contested domain.

As noted in the first two chapters, the lack of cohesive vision and integrated space advocacy at senior levels within DoD is problematic. Emblematic of this challenge has been the absence of a cohesive enterprise architecture for space until the past few years and the persistent lack of synchronization of space assets. DoD has also lacked clarity on how to prioritize space mission sets. In particular, requirements and acquisition communities have perceived a need for exquisite, monolithic solutions for all users, which has affected the agility and responsiveness of space acquisition. Acquisition program success (particularly the success of the expensive satellites) was prioritized over other considerations, including what is considered to be the successful delivery of a synchronized end-to-end space capability. The interviewees agreed that program managers, their staff, and industry have been incentivized to deliver one part of the space capability (an acquisition program) on cost, schedule, and performance, but they are not graded on whether the output of that program has improved the larger space enterprise.

This lack of vertical and horizontal synchronization or alignment of space assets across the space enterprise has been a persistent challenge for DoD and has caused lengthy schedule delays and pieces of capabilities delivered in an uncoordinated fashion. For example, in 2017, the GAO documented that "ground system delays have been so lengthy that satellites sometimes spend years in orbit before key capabilities can be fully used."[1] This finding has followed more than a decade of analysis that identified similar challenges for DoD's space acquisition community.

DoD's governance of the space enterprise is a struggle because of organizations and processes that are separated by many seams. Space requirements have been identified and/or generated by six organizations.[2] Because of multiple other organizational, cultural, and contractual reasons, space assets reside in disparate organizations (and multiple prime contractors).[3] Within the acquisition community there are fragmented acquisition authorities. There has also not been accountability in any one place in DoD for delivering overall space capability. Enterprise architecture performance has been assessed by six or more DoD organizations and agencies, complicated by commercial, civil, and international equities. In addition, funding rules and sources drive an artificial split between pieces of space capability (i.e., satellite, launch services, ground station(s), and user terminals). This has also contributed to a lack of synchronization (i.e., DoD artificially divides the mission into three products or services and expects that integration will succeed when all the pieces have been finally brought together). These problems are not unique to space acquisition. The nuclear enterprise faces a similar disjointed structure within DoD and interagency challenges.[4]

---

[1] GAO, "Space Acquisitions: DOD Continues to Face Challenges of Delayed Delivery of Critical Space Capabilities and Fragmented Leadership," statement of Cristina T. Chaplain, director of Acquisition and Sourcing Management, before the Subcommittee on Strategic Forces, Committee on Armed Services, U.S. Senate, GAO-17-619T, May 17, 2017a, p. 1.

[2] According to the GAO,

> In July 2016, in response to a provision of a Senate Report accompanying a bill for the National Defense Authorization Act for Fiscal Year 2016, we issued a report that reviewed space leadership in more depth and concluded that DOD space leadership was fragmented. We identified approximately 60 stakeholder organizations across DOD, the Executive Office of the President, the Intelligence Community, and civilian agencies. Of these, eight organizations had space acquisition management responsibilities; eleven had oversight responsibilities; and six were involved in setting requirements for defense space programs (GAO, 2019b, p. 10).

These six are AFSPC, the Army Space and Missile Defense Command/Army Forces Strategic Command, the Chief of Naval Operations, the Commandant of the Marine Corps, U.S. Strategic Command, and the Joint Chief of Staff (GAO, 2016b, p. 28).

[3] In the past, this has become an issue, with multiple contractors providing different systems necessary for an overall capability in an uncoordinated fashion, delaying the overall capability potentially beyond the useful life of some components.

[4] As Snyder et al., notes:

> Because the nuclear weapons are owned by the Department of Energy and are highly integrated with the delivery platforms, many of the programs face interagency coordination challenges. Coordination across agency boundaries is exacerbated by separate budgets, different acquisition processes and milestones, and contrasting cultures. Technical difficulties and cost overruns in one agency can imperil an associated program in another. In addition, the Air Force has operated with a fairly constant budget over the past decade, forcing tough decisions on priorities and threatening sustainment of adequate funding levels (Don Snyder, Sherrill Lingel, George Nacouzi, Brian Dolan, Jake McKeon, John Speed Meyers,

Compounding these factors is the need to consider and be cognizant of complex systems working together to provide the desired effects. Having a full understanding of the interdependencies within a system's components helps the decisionmaker understand how decisions affect the system's cost, schedule, and performance. Understanding the complexity inherent with individual systems is a first step—complex relationships among systems also must be understood and managed, because multiple complex systems work together to provide a capability (e.g., missile warning, space situational awareness, satellite communications, and others). Furthermore, these systems interact with and/or depend on each other, and that must be managed also. Thinking about multiple layers of complexity holistically is necessary for enterprise thinking to succeed.

These factors have resulted in a lack of comprehensive and responsive intelligence capability, stovepiped systems, a failure-intolerant culture, and long development cycles. These individual systems have not been integrated, causing a lack of synchronization that had been considered by DoD as a "fact of life," according to interviewees. Significant oversight at all levels (and accompanying acquisition processes) of expensive acquisition programs has also instilled a fear of failure into the workforce, driving it to eliminate all risk before launching space assets, according to interviewees.

It has been reported that peer competitors recognize the strategic advantages of space and are fielding systems to deny them to the United States.[5] U.S. leadership also acknowledges that the development of space capabilities is taking too long and is committed to improving this stage.[6] Accelerating acquisition decisionmaking can help with this improvement.

Rapid Decisionmaking Framing Assumptions

Given the current state of space acquisition, the research team initially identified a set of framing assumptions—a small number of core tenets that, if proven false, will prevent successfully developing and implementing potential approaches for improving space acquisition. As mentioned in Chapters One and Two, the desired characteristics of clean sheet space acquisition are that it be (1) threat-informed, (2) responsive, and (3) cost conscious. From this, the following framing assumptions were identified about rapid decisionmaking and fielding of capabilities:

Kurt Klein, and Thomas Hamilton, *Managing Nuclear Modernization Challenges for the U.S. Air Force: A Mission-Centric Approach*, Santa Monica, Calif.: RAND Corporation, RR-3178-AF, 2019, p. 5).

[5] Defense Intelligence Agency, *Challenges to Security in Space*, Washington, D.C., January 2019; and DoD, *Defense Space Strategy Summary*, Washington, D.C., June 2020a.

[6] Charles Pope, "Raymond and Space Force Enter New, Ambitious Phase as U.S. Space Command Changes," Air Force Public Affairs, August 24, 2020.

- USSF is the originator and developer for DoD space doctrine and concept of operations (CONOP). This fact allows USSF to define its warfighting approach and CONOP from which requirements are developed and policies created. This also provides the guiding principles for decisionmaking and subsequent execution.
- USSF is the final authority for developing, procuring, fielding, operating, and sustaining DoD space assets. With this framing assumption, USSF has the leeway to determine the solutions to its requirements, prioritize their development and distribution, and determine their disposal. However, USSF does not operate in a vacuum and must consider the equities of space enterprise stakeholders. For example, USSF must coordinate with the IC, but the IC has a different mission set and requirements. Coordination with other services—USSF's "customers"—could be similarly or even more challenging, and USSF's ability to deliver will depend, in part, on its ability to provide actionable requirements.
- USSF has the authority to distribute funds received via the budgeting process. Underlying this assumption is the authority for USSF to realign resources to meet its charter of providing space capability for military purposes to the nation in accordance with national security priorities.
- USSF will focus on and prioritize maintaining agreed-on schedules for space acquisition programs. In other words, acquisition programs need to stick to predictable schedules where the actual does not deviate from the estimated in order for synchronization to be successful. This allows for better coordination between the vertical alignment of assets, including satellites, ground stations, and user terminals.

Areas for Improvement

We conducted a literature review on rapid decisionmaking and held discussions with more than 45 acquisition and space SMEs to understand the reasons and potential solutions for a persistently slow delivery of capability, lack of agility and/or responsiveness, stovepiped individual systems, capabilities delivered in an uncoordinated fashion, and lack of overall vision/architecture guiding decisions. These interviews identified several major areas that could be improved to increase the likelihood of implementing rapid, enterprise-centric decisionmaking with faster execution.

Define the Enterprise with an Adaptive Technical Architecture

To make decisions on an enterprise basis, *what the enterprise is* should be defined and promulgated across USSF and its stakeholders. The creation of a technical architecture, based on USSF-developed space warfighting doctrine and CONOPs, would help with that definition. It would also serve as the road map for space operations and interactions with other domains and would provide industry with insight into what USSF sees as its future. However, this technical architecture must be updated constantly with new information, providing off-ramps for systems that prove to be nonviable and on-ramps to insert technology innovations and adapt to changes in the perceived threat.

This technical architecture provides the framework within which space acquisition decisions are made. From the architecture, acquisition decisionmakers derive USSF's capability priorities and use them to drive the acquisition process.

USSF should also use this architecture to promote industry investment to achieve the desired results. However, as mentioned previously, the architecture must be adaptable to environmental changes—threat and industry innovations. A static space architecture rapidly becomes an irrelevant one.

Manage Space as a Portfolio to Deliver Capability

DoD's space portfolio is a complex system of systems. USSF must be allowed to govern and manage space as a portfolio rather than as individual acquisition programs; this will allow it to balance risks, allocate resources to higher priority activities, and ensure that complete capabilities are delivered according to predicted schedules. However, this approach requires significant insight into all aspects of the portfolio so decisionmakers can understand the necessary trade-offs and make data-driven decisions, including mapping out the technical architecture and how the space system of systems work together.[7]

Developing and establishing a technical architecture would provide a guide to anyone building components of space assets. This would enable them to understand how these components integrate into the larger enterprise and their importance within the larger mission context. The architecture would allow USSF leadership to oversee and execute space acquisition holistically, providing capabilities and effects rather than individual acquisition programs or systems. Transitioning from program managers to capability managers would be a step in the right direction. Capability managers would own the whole end-to-end delivery of space systems providing (e.g., tactical SATCOM; weather; positioning, navigation, and timing; and others).

To facilitate the delivery of capability, USSF should consider alternatives to traditional monolithic systems. Reallocating the scores of requirements in these higher-risk programs to smaller systems would enable shorter schedules. Shorter schedules, or schedule increments, provide the capability managers with increased progress feedback, which, in turn, allows for more-rapid adjustments—responsiveness and agility—and increased management attention to maintain agreed-on timelines. A refocus on smaller systems could also contribute to program synchronization within the portfolio.

Development of common standards, facilitated and agreed to by industry, could be a way to enhance capability delivery. Common interface standards for ground stations, user termi-

[7] There is a rich literature on systems of systems in systems engineering. For example, see MITRE, "Systems Engineering Guide: Systems of Systems," webpage, undated; and Purdue University, College of Engineering, "System of Systems (SoS)," webpage, undated.

nals, and classes of buses could, for example, promote opportunities for interchangeability and increase flexibility where it makes technical sense. Thus, common standards would give capability managers more options for staying on schedule. However, common interface standards can constrain future innovation and evolution and must be used judiciously.

Increase Funding Flexibility to Promote Execution Agility

Although the acquisition community needs to be agile, some acquisition decisions are constrained by laws, regulations, and policies. An example of such a constraint is the Financial Management Regulation, which was instituted to ensure that public funds are used and accounted for appropriately. Within its more than 7,000 pages, it directs proper uses for different appropriations and management by program elements (PEs).[8] Although no one disputes the need for accountability, it can make members of the acquisition community extremely cautious in order to avoid any potential violations of any laws, policies, or regulations. This level of caution has contributed to and perpetuated a culture of risk averseness. According to some interviewees, this culture of risk averseness has driven decisionmakers to continually question the sufficiency of compliance and to request more data, searching for the perfect amount and type of information to support a decision, and thereby continually delaying it.

With the PE being the foundation of DoD's planning, programming, budgeting, and execution process (and each PE being associated with a specific activity), decisionmakers have traditionally focused on those activities individually. Potential consequences of this focus are individually optimized, stovepiped systems with reduced interoperability and with hardware and software delivery optimized for the system instead of the overall capability.

One way to increase flexibility (while remaining in compliance with existing laws, regulations, and policies) would be to reduce the number of PEs the maximum extent possible. Maximum flexibility would be one PE for all USSF acquisition. The acquisition decisionmaker could address priorities quickly. However, this could be difficult to achieve because it would require a significant change in how congressional oversight is maintained. Therefore, an intermediate approach to increasing flexibility would be consolidating into capability PEs.[9] Another would be eliminating "colors of money" (e.g., research, development, test, and evaluation; procurement; and operation and maintenance). Such policy changes would be especially relevant to space because its programs do not typically align with the traditional major capability acquisition process phases.[10] Reprogramming flexibility could be used to

---

[8] Office of the Under Secretary of Defense (Comptroller), Chief Financial Officer, *Department of Defense Financial Management Regulation (DoD FMR)*, Washington, D.C., DoD 7000.14-R, May 2019.

[9] This should be done as much as possible within the authorities of USSF. Because these are consolidated, USSF will need to establish data governance focused on data collection, storage, transparency, and ease of sharing. This would allow USSF to provide Congress and other stakeholders visibility into the allocation and expenditure of public funds.

[10] Many space acquisition systems do not align with the traditional major capability acquisition, sometimes referred to as the *waterfall* approach (as described in Defense Acquisition University, "DAU Adaptive

move funding between space assets in the space mission portfolio to fix synchronization and other issues, thus providing the capability manager the flexibility to rapidly allocate resources to higher-priority activities and help eliminate the artificial barriers between space assets that were created by the funding issues.

Although the PE structure is part of what allows Congress insight into (and oversight of) executive branch investments, this is not the only approach into ensuring that necessary transparency. Increased transparency from USSF through enhanced data-sharing, insight via performance metrics, and more frequent and open interactions with Congress could be a viable substitute. USSF should continue this outreach to establish a positive and trusted relationship with Congress.

This combination of changes to the existing budget structure, described earlier, allows USSF leadership to consider space as an interrelated portfolio of capabilities rather than as a bunch of unrelated programs with separate PEs.

Establish a Single Space Acquisition Decisionmaker

Discussions with experienced space acquirers from multiple organizations revealed that while many individuals make decisions within space acquisition that affect program execution,[11] it can be difficult to find, much less access, the right individual in a timely manner. Layers of formal (and informal) management have led to and continue to exacerbate this situation.

The FY 2020 NDAA established the Space Force Acquisition Council (SAC). Its membership includes the Under Secretary of the Air Force; Assistant Secretary of the Air Force for Space Acquisition and Integration (who also acts as SAC Chair); Assistant Secretary of Defense for Space Policy; NRO director; CSO; and Commander, U.S. Space Command (CDR USSPACECOM). The SAC "shall oversee, direct, and manage acquisition and integration of the Air Force for space systems and programs to ensure integration across the national security space enterprise."[12] Although it is clear that the intent of the SAC is to improve the state of space acquisition, evidence shows that numerous other attempts at decisionmaking by committee have actually done the opposite.

Acquisition Framework," webpage, undated-a). Many space programs fly the hardware created in the traditional Engineering and Manufacturing Development (EMD) Phase. Also, there is no traditional production run. Therefore, being required to use research and development (R&D) funds in EMD and procurement funds in Production requires a significant amount of creativity in the program office to comply with the Financial Management Regulation.

[11] The research team interviewed numerous space acquisition experts, including the Assistant Secretary of the Air Force (Acquisition, Technology and Logistics) and officials from National Reconnaissance Office (NRO), Space Rapid Capabilities Office (SpRCO), Space Development Agency (SDA), and Space and Missile Systems Center (SMC). Many recounted difficulties that they had personally experienced either getting to someone who could make the necessary decision or being able to gather authoritative information to make a decision in a timely manner.

[12] See Section 954, Space Force Acquisition Council of the FY 2020 NDAA (Pub. L. 116-92, 2019).

- The Reagan administration created an advisory group within the National Security Council. The Senior Interagency Group (SIG) for Space (SIG[Space]) members did not report directly to the U.S. President. SIG(Space) also did not have decisionmaking authority, which resulted in little results and many turf battles. Congress viewed the SIG(Space) as unproductive.
- In the first Bush administration, the National Space Council had high-level membership and a more direct line to the President. The National Space Council was intended as a means of controlling space policy between the National Aeronautics and Space Administration (NASA), Congress, and the President. Early actions were positive because the National Space Council saved two important programs. However, the Space Exploration Initiative in July 1989 diverted attention from other space policy matters. The National Space Council began to also experience turf battles.
- In the second Bush administration, the Space Policy Coordinating Committees tried to involve all stakeholders in the executive branch. This inclusiveness led to inefficiency. Senior-level members were not involved until the end of the process, which led to issues that could have been prevented by these members' early input.[13]

The optimal decisionmaking approach would follow the concept of "unity of command"—a military principle to ensure "unity of effort under one responsible commander for every objective."[14] Therefore, the optimal approach for space acquisition is a decisionmaker vested with the necessary insight, oversight, and authorities. This single decisionmaker would be responsible for acquisition budgeting, developing, and implementing technical solutions to fill approved capability gaps and executing the Defense Acquisition System (DAS). This Head of Space Acquisition (HSA) would report to USSF leadership—the Secretary of the Air Force and the CSO—and oversee the space acquisition enterprise as depicted in Figure 3.1, ensuring that it is threat-informed, responsive, and cost conscious. A Space Acquisition Board's membership would consist of the HSA (Chair) with the Secretary of the Air Force, CSO, and CDR USSPACECOM, the leaders controlling USSF's execution of the DAS, resources, and requirements, respectively. These are all elements needed to effectively execute space acquisition.

To accelerate execution, the HSA could default to delegating decisions to the lowest possible level, with elevations occurring only in extreme circumstances.[15] Empowered capability managers would be responsible for an end-to-end capability (e.g., combination of satellite, ground station, and user terminals) to reduce seams between programs and increase the like-

[13] James A. Vedda, *Center for Space Policy and Strategy Policy Paper, National Space Council: History and Potential*, El Segundo, Calif.: Aerospace Corporation, November 2016.

[14] Joint Publication 3-0, *Joint Operations*, January 17, 2017, incorporating change 1, Appendix A, *Principles of Joint Operations,* October 22, 2018.

[15] Interviewees agreed, pointing out that significant time is lost in the staffing process to get decisions.

FIGURE 3.1
Proposed Acquisition Governance Organization

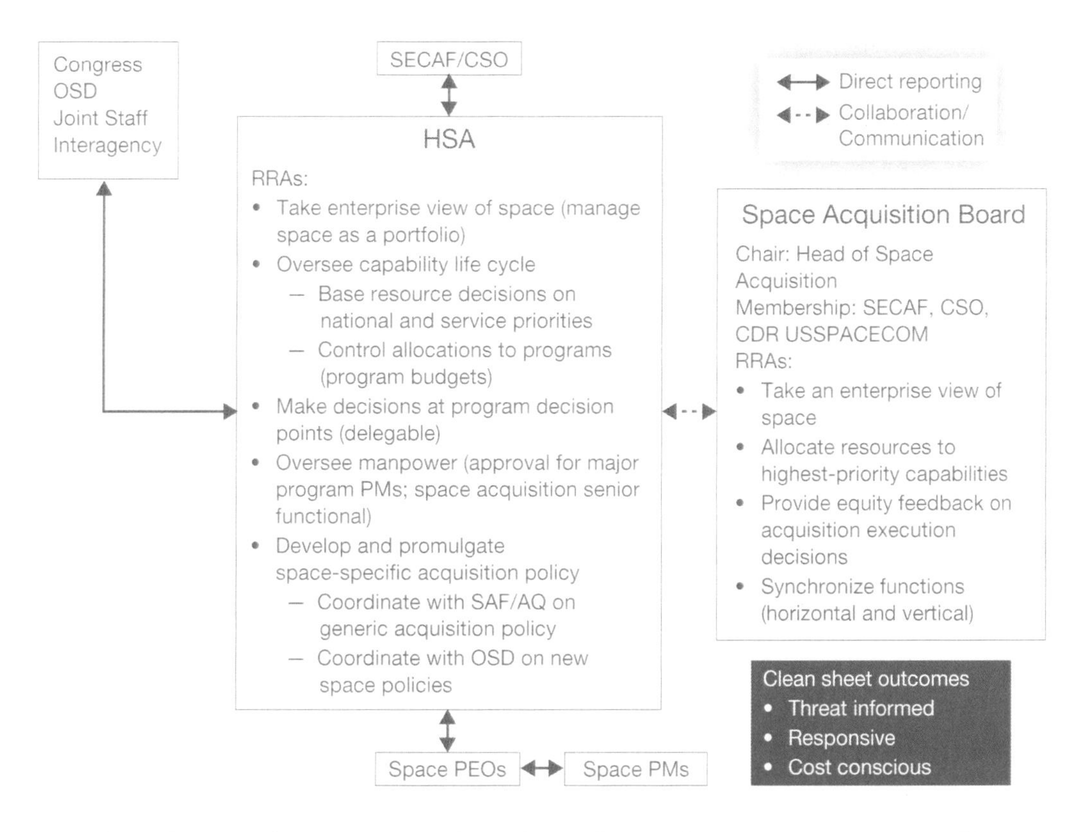

NOTE: OSD = Office of the Secretary of Defense; PEO = Program Executive Officer; PM = program manager; RRAs = roles, responsibilities, and authorities; SAF/AQ = Assistant Secretary of the Air Force (Acquisition, Technology, and Logistics); SECAF = Secretary of the Air Force.

lihood of on-time capability delivery. However, for this approach to be successful, a talented, experienced workforce would be necessary.

Develop an Effective Workforce

Among the most critical elements needed to implement effective space acquisition are those involving the workforce actually executing the DAS.[16] USSF should incentivize the acquisition workforce to deliver enterprise capability and to radically change its risk posture.

A robust systems engineering capability within USSF, woven into the acquisition process, could help focus the workforce on capability instead of individual systems (e.g., similar to the NRO's mission integration directorate). In addition, a workforce highly experienced in

[16] Specifics regarding the changes needed for the space acquisition workforce are described in Chapter Four.

the use of the newest techniques in digital engineering[17] (the preferred prototyping method) could help use resources more efficiently, decrease development time, and assume a more-tolerant approach to risk.[18]

Focus on Delivering Complete Capability on Time

USSF's biggest priority will probably be adapting the delivery of new capabilities to the ever-shrinking timelines engendered by the threat. Some potential methods to maintain schedule include the following:

- prevent requirements creep
- define what is necessary for the warfighter in the near term and build on incrementally over time
- use prototypes to burn down technical risk early (Congress has provided additional support for rapid prototyping in the Middle Tier of Acquisition pathway)
- use a Modular Open Systems Approach (i.e., a technical and business strategy for designing an affordable and adaptable system)
- incentivize contractors to maintain schedule using contracts incentives.

This adaptation includes building capability through predictable, time-phased increments that prioritize urgent needs and deliver those incremental capabilities first using such techniques as the Modular Open Systems Approach. Likewise, incremental capability delivery and predictable scheduling allows better coordination between the vertical alignment of assets, including satellites, ground stations, and user terminals. Although difficult, given the changing threat, USSF's acquisition programs should stick to predictable schedules where the actual does not deviate from the estimated. This increases the likelihood that synchronization will be successful (potential approaches provided below); otherwise, USSF will continue to face the challenges as documented by the GAO over the past decade (see Figure 3.2). For example, in 2009, the GAO found that for eight major acquisition programs, three of the satellites were estimated to be delivered well before ground systems, and there was a similar problem with user terminals.[19] This same problem persisted over time and was again identified in the 2017 GAO study.[20] Likewise, the GAO also discussed the management problems

[17] According to the Defense Acquisition University, *digital engineering* is "an integrated digital approach that uses authoritative sources of systems' data and models as a continuum across disciplines to support lifecycle activities from concept through disposal" (Philomena Zimmerman and Darren Rhyne, "Lunch and Learn—Digital Engineering," presentation, Defense Acquisition University, May 23, 2018).

[18] Some of the benefits of this increased tolerance to risk in industry interactions are highlighted in Chapter Six.

[19] GAO, *Defense Acquisitions: Challenges in Aligning Space System Components*, GAO-10-55, October 2009.

[20] GAO, 2017a.

FIGURE 3.2

U.S. Government Accountability Office Repeatedly Identified Synchronization as a Challenge for Space Systems

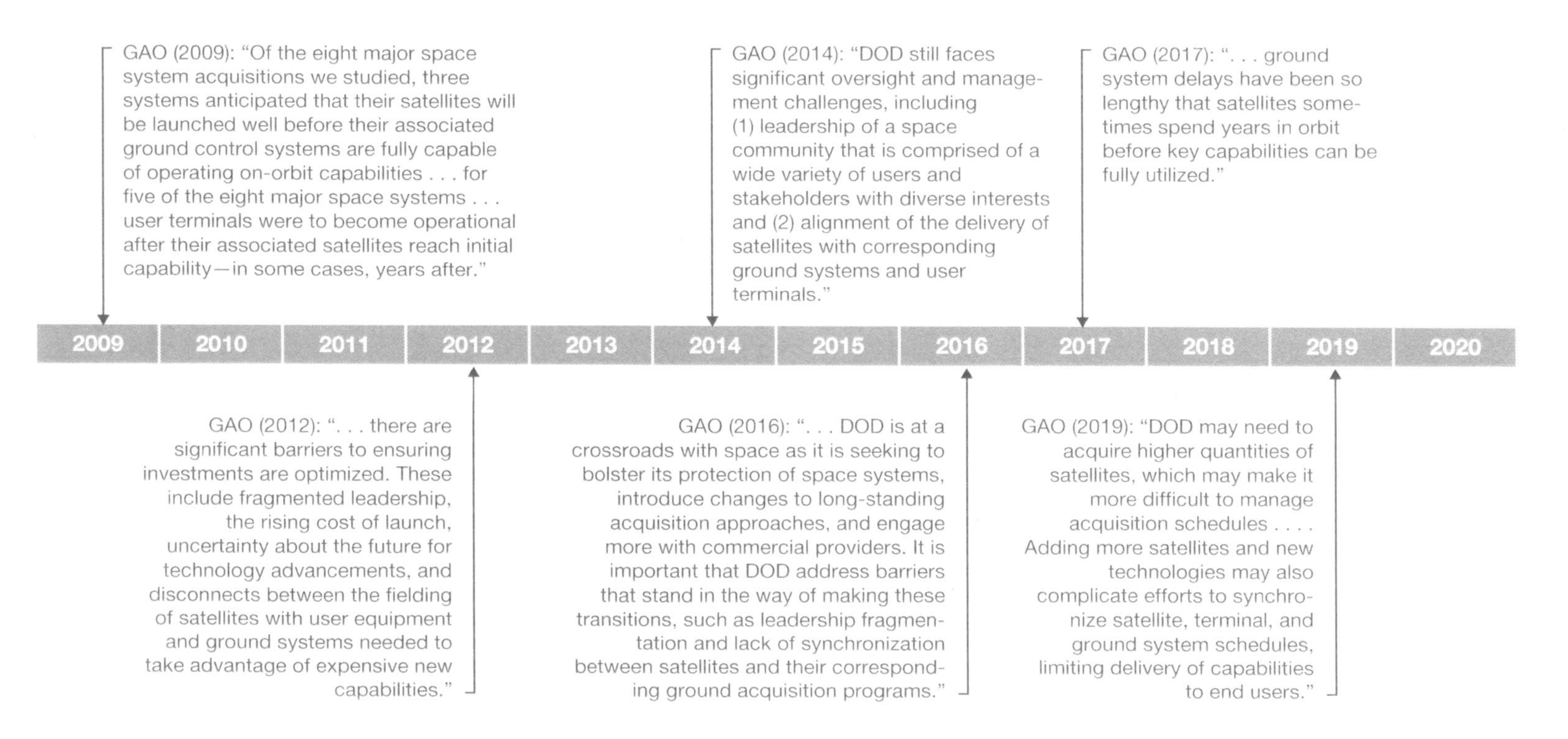

SOURCES: GAO, 2009; GAO, "Space Acquisitions: DOD Faces Challenges in Fully Realizing Benefits of Satellite Acquisition Improvements," statement of Cristina T. Chaplain, director of Acquisition and Sourcing Management, before the Subcommittee on Strategic Forces, Committee on Armed Services, U.S. Senate, GAO-12-563T, March 21, 2012; GAO, "Space Acquisitions: Acquisition Management Continues to Improve but Challenges Persist for Current and Future Programs," statement of Cristina T. Chaplain, director of Acquisition and Sourcing Management, before the Subcommittee on Strategic Forces, Committee on Armed Services, U.S. Senate, GAO-14-382T, March 12, 2014a; GAO, "Space Acquisitions: Challenges Facing DOD as It Changes Approaches to Space Acquisitions," statement of Cristina T. Chaplain, director of Acquisition and Sourcing Management, before the Subcommittee on Strategic Forces, Committee on Armed Services, U.S. Senate, GAO-16-471T, March 9, 2016a; GAO, 2017a; GAO, 2019c.

for leadership with these space programs over time, such as fragmentation through a wide variety of users and stakeholders of these systems.

Likewise, identifying upfront what delivering a successful space capability means is also important: For example, do we need a global solution now, or can we add capability over time and prioritize the most important users? As part of defining the capability, it would be helpful for the acquisition and user communities to agree on synchronized milestones and credible initial operational capability (IOC) date. An institutionalized feedback loop with warfighters would help to improve synchronization and to do rapid upgrades.

Considering space assets at the portfolio level will also help maintain schedules, as will the use of predictable schedule increments. With a top-level view of the entire portfolio, leadership will be able to make more-informed decisions on nontraditional alternatives versus traditional monolithic, multipurpose systems. Considering assets at the portfolio level will also help to visualize and achieve a balance between higher risk and lower risk investments.

Anchor Milestones; Shorten Durations Between Milestones

To make sure that USSF requirements and acquisition communities are addressing the most important needs based on a schedule that coincides with the threat, USSF should consider "anchor" delivery milestones. These are priority capabilities that need to be in the delivery, in contrast with others that are lower priority and could be delivered later or dropped. This could potentially require prioritizing schedule over other requirements.

Foster Synchronization by Increasing Collaboration and Communication

Acquisition strategies could be used to focus on synchronization of contracts, using joint performance objectives to incentivize synchronization by the prime contractors among space assets, and setting expectations among all stakeholders. Acquisition strategies also could be tailored to potential risks for each program and also to define the IOC as the date when the satellite, ground station, and user terminals are needed together.

Communication and transparency are critical to improving rapid decisionmaking and fostering an enterprise view of space capabilities. Understanding potential technical and other challenges across all stakeholders will help align important timing (e.g., launch dates).

Summary

Basing recommendations on the totality of these findings could provide USSF acquisition process with the ability to manage space as an enterprise with agile decisionmaking, focusing on the nation's and USSF's highest priorities. These efforts will help promote the coordinated delivery of capabilities and help the United States remain ahead of potential adversaries.

CHAPTER FOUR

Talent Management

Understanding the Importance of Talent as an Asset

Talent is an organization's best and most important asset. The standing up of USSF is an opportunity to create not only a new unified identity and culture, but also to reorient all military and civilian personnel as space "warrior engineers." To achieve a culture where space acquisition is a warfighting capability and the United States leads in space superiority, everyone must believe and work to meet the mission as a unified force. Incentives must be aligned with desired outcomes across the enterprise. This chapter discusses the importance of true independence as a service to instantiate the desired culture, along with paths forward to realize USSF goals through transforming talent management.

Talent Management Framing Assumptions

Several framing assumptions underlie our approach to talent management in the clean sheet vision for USSF space acquisition. These assumptions are the minimum needed to organize, train, and equip space professionals to meet this vision of space acquisition as a warfighting capability. The discussion throughout this chapter highlights the importance of each of these assumptions:

- **Assured access to the required talent:** USSF must have continual access to the talent it needs to be able to effectively manage the force to achieve its vision. Access to talent applies to both uniformed and civilian personnel.
- **Clear guidance from leadership and chain of command to reinforce the vision:** USSF leadership must embrace the clean sheet vision and foster buy-in at all levels of the USSF enterprise. Leadership must clearly vocalize the clean sheet vision as the desired end-state for USSF, creating and enabling a work environment that allows space professionals to operate as needed to achieve the vision.
- **Needed legislative or policy changes that are achievable:** Should any of the recommendations require formal revisions to existing policies or legislation, we assume these changes are actionable within a reasonable time frame.

- **Culture and incentives in place to align behavior:** Alongside support from leadership, policies and practices are implemented that enable effective talent management.

Without these assumptions in place, the ability for USSF to implement many of these recommendations could be difficult, if not impossible.

Importance of Identity and Culture to Talent

A large body of business literature focuses the interconnection and reinforcing nature of brand, identity, and culture and the importance of these factors in recruiting and retaining top talent.[1] USSF, as a new organization, is no different. Through building a culture and identity, USSF can create its desired work environment and motivate its personnel to accomplish the mission rapidly and effectively.

Consider USSF's primary goal, although still nascent, of fostering an identity and culture that will grow and cultivate top talent as part of an engaged, motivated, and effective workforce.[2] This culture, discussed in more detail later in this chapter and in Chapter Six, will enable decisionmaking that prioritizes delivering enterprise capability and radically changes the enterprise's approach to risk. This new approach could reward out-of-the-box thinking and creative solutions while tolerating the potential failures of rapid innovation. It could also nurture and reward strong, data-driven judgment and excellence in leadership and team management. Finally, this new approach encourages collaboration with traditional and nontraditional players in the commercial space industry, as we will discuss further in Chapter Five.

Transformation in Space Identity and Mission

The creation of USSF provides an opportunity to build a unified space identity and culture, one aspect of which is space acquisition as a warfighting capability. This could be a next iteration of SMC as USSF stands up SSC, using lessons learned, along with industry best practices on culture and brand-building.

There are benefits to attracting and retaining top talent from creating a revitalized space identity and mission that defines space acquisition as a warfighting capability. This redefined identity can jumpstart projecting a culture to the world that

- improves USSF recognition
- institutionalizes space acquisition as a warfighting capability

---

1 ZipRecruiter, "Why Company Culture Is So Important for Attracting Talent," blog post, March 25, 2015.

2 Susan Milligan, "Use Your Company's Brand to Find the Best Hires," *HR Magazine*, September 3, 2019.

- creates trust in USSF and the ability to meet its mission
- builds a space community through a unified identity
- sustains a pipeline of attracting and retaining top talent
- motivates and empowers the space workforce.[3]

USSF culture, regardless of the shape it takes, will be embedded within every aspect of the enterprise and its talent. This means that to successfully foster the desired identity and culture, efforts will need to be multidimensional, continual, iterative, and reinforcing. For example, transforming space acquisition processes will transform the beliefs and norms of the space acquisition workforce. At the same time, transforming how this talent is managed will transform how the space acquisition workforce approaches each step of a space acquisition process and program.

U.S. Space Force Transformation Requires Complete Independence

At the core of building a new identity of space acquisition as a warfighting capability, USSF must be truly independent. Raymond, the CSO, has explicitly acknowledged the need for USSF independence.[4]

The need for independence is twofold: (1) for freedom to innovate to maintain U.S. superiority in 21st century space and (2) to foster internal and external stakeholder support for USSF and its redefined mission, along with the resources to accomplish the mission. Freedom to innovate allows USSF to transform space acquisition and operations, including talent management. Generating excitement for the future of space, as a separate warfighting domain that is a peer to other services, will elevate its importance to decisionmakers, industry, and top talent. This will help to increase the support from stakeholders from top talent and congressional staffers. And this top-level involvement, in turn, will help to encourage external stakeholders to provide the needed resources and flexibilities to achieve USSF's goals. A credibly independent identity will also compel internal leaders and the space workforce to refocus priorities in line with USSF's redefined mission.[5] For many employees, believing in the mission is an important part of retention. USSF has an incredible branding and engagement opportunity here as it stands up a service dedicated to space. We recommend fully embracing and capitalizing on this moment.

---

[3] Independent Business Association, "The Importance of Business Branding," *Medium*, February 11, 2018.

[4] David Vergun, "Space Force Leader Discusses Newest Military Service," *DoD News*, October 27, 2020.

[5] Natalie Baumgartner, "Build a Culture That Aligns with People's Values," *Harvard Business Review*, April 8, 2020.

In the quest for analogies of independence, the Marine Corps has been routinely touted as an exemplar for USSF.[6] This is because the Marine Corps, even though it is technically part of the U.S. Department of the Navy (DoN), has a very distinct brand, identity, and culture.[7] The Marine Corps, through this independence, has fostered an identity and culture that effectively attracts and retains the necessary talent.[8] At the same time, the Marine Corps leverages some resources and capabilities from the DoN.

However, particularly with regard to recruiting and managing talent, the Marine Corps is a far more independent service than the current USSF. Armed with the uniquely Marine Corps brand, the service has its own, completely separate recruiting apparatus. Marine recruiters are out in strong force across the country, visiting high schools and having separate recruiting offices.[9]

In contrast, USSF is initially relying on the recruitment infrastructure of the U.S. Air Force. For officers, accessions still come from the U.S. Air Force Academy, Air Force Reserve Officer Training Corps programs, and Officer Training School. For enlisted personnel, USSF uses the Air Force Recruiting Service (AFRS) apparatus, using U.S. Air Force recruiters along with AFRS marketing, lead generation, and outreach services. For civilians, recruitment, hiring, and other personnel processing continues through the Air Force Civilian Service (AFCS).[10]

This situation creates a risk that USSF will not have guaranteed access to the most-qualified talent and creates confusion for prospective talent. For example, the "Space Careers" website is potentially confusing for interested military and civilian personnel. When one goes to the USSF homepage and clicks on "Careers," users are not directed to a USSF-hosted careers page but instead are redirected to a U.S. Air Force careers landing page. (Note that this was the case when this research was conducted: The approach for this and the issues highlighted below might evolve over time.)

---

[6] Vergun, 2020.

[7] Dean Crutchfield, "Happy Birthday to America's Most Enduring Brand: The Marines," *Forbes*, November 14, 2011.

[8] The Marines consistently meet annual recruiting goals. See, for example, a discussion of the strength of their brand, outreach campaign, and recruiting force in meeting mission: Shawn Snow, "The Corps Is Finding New Marines Despite Recruiting Challenges," *Marine Times*, November 2018; and U.S. Department of Defense, "Face of Defense: Recruiting the Next Generation," October 2020b.

[9] The U.S. Air Force has the fewest number of recruiters of any service. As a result, and given recent years' recruitment quotas as set by the NDAA, on average recruiters' monthly production goal is 2.1 recruits, much higher than that of other services (see Stephen Losey, "Air Force Aims to Modernize Recruiting Amid Growing Challenges," *Air Force Times*, November 2, 2018). Although progress is being made, U.S. Air Force recruiters have routinely struggled with being overworked and undermanned. See, for example, Stephen Losey, "Crushing Demands of Job Lead Some Air Force Recruits to Falsify Reports," *Air Force Times*, September 1, 2014.

[10] Even for civilian hiring done through USSF direct hiring authorities, candidate onboarding must be processed through AFCS.

Figure 4.1 depicts the top search result and associated webpage for the internet query "space force careers." As just described, the landing page for space careers brings the user to the U.S. Air Force careers page. Only after the user scrolls down are there options for space-related career fields.

Figure 4.2 depicts a potentially more-confusing webpage: the official USSF "Careers" page. This is different from the page in Figure 4.1, which is the top internet search result. This is the landing page for users who go to the official USSF website and click on "Careers" in the top menu. It redirects to a U.S. Air Force website that is seemingly co-branded, which could

FIGURE 4.1
The Search-Based U.S. Space Force Careers Website Is a U.S. Air Force–Branded and –Administered Webpage

SOURCE: U.S. Air Force, "Air Force Careers: Explore Careers and Find Your Purpose," webpage, undated-a.

FIGURE 4.2
The Official USSF Careers Website Is a U.S. Air Force–Branded and –Administered Webpage

SOURCE: U.S. Air Force, "The Sky Is Not the Limit," webpage, undated-b.

raise questions from interested recruits. Moreover, it is unclear why both approaches to learning about USSF careers can bring someone to two different webpages.

Similarly, as noted, civilian hiring remains through the Air Force Personnel Center (AFPC)'s AFCS. Having to work through this additional level of authorities, outside USSF, could result in talent issues far greater than confusing interested talent. Beyond talent issues like delayed hiring, being tied to U.S. Air Force processes limits USSF's ability to create new personnel policies that promote alignment to desired space-specific performance. USSF needs both its own hiring priorities and authorities.

Figure 4.3 shows one example of a LinkedIn recruitment advertisement for civilian USSF positions. The post suggests that the U.S. Air Force is presenting these options, supporting USSF but not coming from USSF itself. Such messaging could make it difficult for prospec-

FIGURE 4.3
Air Force Civilian Service Advertisement of Civilian Space Force Careers on LinkedIn

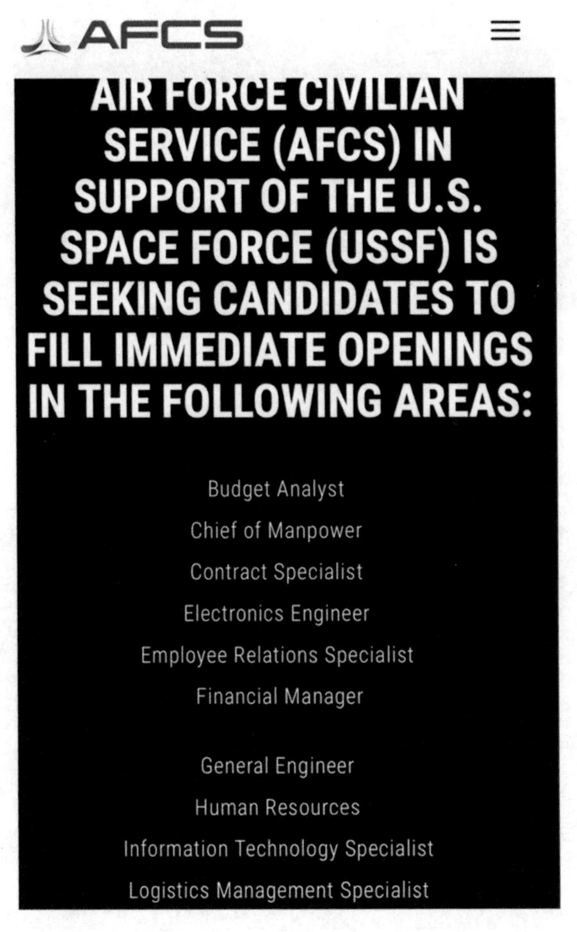

SOURCE: U.S. Air Force, "The Sky Is Not the Limit," webpage, undated-b.

tive applicants to get excited about and engaged with a uniquely USSF mission, because such advertising lacks a clear USSF identity and culture.

USSF will have a smaller uniformed presence than the Marine Corps does, particularly on the enlisted side. However, this smaller size does not negate the need for USSF to have its own identity and brand. One risk is that USSF may not be seen as something truly new, merely "AFSPC renamed,"[11] or that it may be confused with other parts of the space community, such as NASA. Exacerbating this concern is that the U.S. Army, for example, is also posting job advertisements on such sites as LinkedIn for its own space operators.[12] Given the fragmented and uncoordinated legacy of the space acquisition workforce,[13] USSF now has an opportunity to unify it and bring it to bear on space acquisition.[14]

Transformation from Airmen to Space Warrior Engineers

Without complete independence, USSF will most likely struggle to realize an optimal talent management strategy that aligns identity and brand to desired culture and performance. A full transformation in talent management and human capital planning would be needed. For example, this overhaul would involve clearly defining desired performance outcomes that are closely tied to building the desired culture and identity. We note upfront that this would be a continual process requiring careful planning and integration of workforce strategy with operational doctrine and strategies.[15]

The effort to unify the space cadre in this direction began in 2016, with Gen John Hyten's vision for a Space Mission Force. This vision directed service leaders in AFSPC to instill uniform mission goals and training across the U.S. Air Force's space military component.[16]

USSF has the opportunity to continue to build on this vision by unifying the entire space cadre—military and civilian—as space warrior engineers. USSF recently announced that its space professionals will be known as "Guardians." An entire culture should be built around this identity, actively cultivating what it means to be a Guardian and the mission, ideals,

[11] Air Force Space Command (AFSPC) was established in 1982 as a major command with space operations as its primary mission (see Air Force Space Command, "Air Force Space Command History," webpage, undated).

[12] LinkedIn.com, "Already a Space Pro? Add Army Space Ops Officer to Your Resume," GoArmy promotion, October 12, 2020. This Army promotion advertisement links directly to the U.S. Army Talent Management website (U.S. Army Talent Management, homepage, undated).

[13] GAO, *Defense Space Systems: DOD Should Collect and Maintain Data on Its Space Acquisition Workforce*, GAO-19-240, March 2019a.

[14] GAO, 2016b.

[15] Society for Human Resource Management, "Aligning Workforce Strategies with Business Objectives," September 2015.

[16] John E. Hyten, *Space Mission Force: Developing Space Warfighters for Tomorrow*, Air Force Space Command, white paper, June 29, 2016.

and values it represents. This could be analogous to the culture built around *wingmanship* in the U.S. Air Force, which refers to the entire community of military members, civilians, and contractors as "wingmen." All USSF service members, uniformed and civilian, will be Guardians, although many civilians will remain technically employed by the U.S. Air Force.

In shaping what it means to be a Guardian, we recommend that the role should be focused on the vision of being an agile and adaptable "warrior engineer." The vision for USSF to be small and lean should be embedded into this concept as an enabler and an asset, in which Guardians leverage the resulting agility to their advantage. USSF should have fewer bureaucratic layers, and decisions can be delegated to the lowest level. Combined with requested changes in space acquisition funding authorities and procedures, recent industry innovation in commercial space satellite design and functionality, and concurrent modernization of satellite architectures as multiorbit and proliferated, the stage has been set for acquisition to happen more rapidly and iteratively. Modular satellite design allows for iteration; more and cheaper satellites allow for more prototyping. The resulting agility affords Guardians the opportunity to try innovative and untested ideas and the flexibility to reward rapid decision-making and risk tolerance that was previously untenable for many junior officers overseeing these large programs.

However, a Guardian identity that leverages the advantage of a smaller USSF relies on having both the right talent and the right management of that talent; this identity would align incentives to desired performance and create career pathways that excite and engage top talent. This includes avenues for growing and cultivating talent and for enhancing the permeability of talent to rotate in and out of service. Providing multiple on- and off-ramps creates more options for top talent that is willing and able to serve, but on shorter or intermittent terms.

In addition to being the smallest service, USSF is distinct in that civilians will likely outnumber uniformed personnel. This imbalance is due to how talent was organized in AFSPC, which was later transitioned as the initial core component of USSF. Space acquisition programs, like many defense acquisition programs, are long and costly (as noted in Chapter Three). Given that military personnel rotate assignments every two to three years, maintaining a consistent officer presence throughout an acquisition life cycle was not feasible, given the current views regarding "homesteading" and its impact on active-duty promotions. Instead, civilians have been relied on to provide consistency to the program team, performing tasks similar to uniformed personnel potentially staying on board for much of the life of the program. Civilians need to be a core part of any transformation in brand, identity, and culture.

Interviews with stakeholders and SMEs across the space community provided insights for how USSF can approach talent management to build a unified culture of space warrior engineers that embody space acquisition as a warfighting capability. We discuss each in turn.

Treatment of and Expectations for Civilians and Uniformed Personnel

The success of any organization's mission depends on the success of its talent. Our interviewees repeatedly emphasized that they did not delineate roles and responsibilities between civilians and uniformed personnel because of the reliance on civilians in space acquisitions. Both worked together to realize the program and provide capability to operations.

Regardless of whether they were military or civilian, SMEs we spoke with noted that space warrior engineers must have advanced technical skills, ongoing familiarity with industry advancements, perspective on space as a warfighting domain, excellence in management and leadership, and strong motivation inspired by the mission. They must be agile, data-driven, risk-tolerant, innovative problem-solvers to outpace unexpected threats. Most of these qualities should be expected in all space professionals, whether or not they are on the frontlines of acquisition or satellite operations.

A common theme followed: to treat civilians and military personnel similarly—in assignment, performance evaluation and management, career advancement and promotion opportunities, and incentive-related compensation. Equal treatment will facilitate aligning all uniformed and civilian advancement policies with USSF's desired performance goals.

Additionally, interviewees noted that FFRDCs provide significant technical expertise across programs, and systems engineering and technical assistance contractors provide valuable systems engineering and other support. Many were even colocated with military and civilian personnel. FFRDC SMEs reported to their technical manager, who assigned them to programs for certain phases of the acquisition process or to provide technical support as needed. Although more assessment is needed to fully identify how to best leverage and organize this talent in a new USSF, this segment of the space workforce could also benefit from being considered part of space enterprise as warrior engineers.

Bring All Personnel Operations in House

USSF vision is to ensure U.S. leadership in space-based innovation, secure the United States' space-based capabilities, and develop world-class joint warfighters to protect and defend the United States.[17] According to Raymond, the CSO:

> We are forging a warfighting Service that is always above. Our purpose is to promote security, assure allies and partners, and deter aggressors by demonstrating the capability to deny their objectives and impose costs upon them. We will ensure American leadership in an ongoing revolution of operations in space, and we will be leaders within government to achieve greater speed in decision-making and action. We will partner with and lead others to further responsible actions in, and use of, space to promote security and

[17] Raymond, 2020.

> enhance prosperity. Should an aggressor threaten our interests, America's space professionals stand ready to fight and win.[18]

Therefore, USSF needs the flexibility afforded by its own talent management system that will contribute to the fight by effectively managing the service's most valuable resource through its recruitment, training, development, and retention strategies.

Instead of USSF relying on the AFPC, interviewees noted the advantage of building one in house to maintain independence. Doing so would serve to establish an independent USSF employee personnel system designed around USSF's own brand and mission. Such efforts could include expanding the scope of STARCOM to become USSF's version of AFPC in addition to current plans for training coordinator (e.g., Space Force's Air Education and Training Command). STARCOM should work alongside the USSF Chief Human Capital Officer to build its own human capital strategy and talent management infrastructure from end to end. It could include USSF-specific performance evaluation, industry partnerships and rotations, functional communities, training, direct hiring authorities as needed to recruit top talent, and, as needed, working with the U.S. Air Force to provide training and education that USSF is unable to feasibly provide.

Implement Selective Entrance Criteria

Growing and cultivating the best talent requires being selective. USSF should create a reputation of prestige to serve, where entry is competitive, similar to other selectively manned organizations. Selectivity will further reinforce identity and culture, as will ensuring that USSF is attracting talent truly motivated by the mission and role. Working for USSF should be a career enhancement, whether on a temporary detail, on a multiyear joint assignment, or as part of a full military or civilian career.

Interviewees and discussions with SMEs supported the idea of implementing selective entrance criteria as a way to maintain high standards. When they were asked about the perceived success of such organizations as NRO and the SpRCO, they repeatedly discussed selective criteria as a key factor. Having the choice of talent and knowing the desired knowledge, skills, abilities, and other attributes of effective personnel has helped ensure these organizations were getting the best and brightest.

Selective criteria should apply to all USSF personnel: enlisted, officers, and civilians entering the service and personnel from other services rotating in on assignment. Such criteria, although not within the scope of this study, should reflect the characteristics of space personnel USSF seeks to cultivate, as noted above. In discussions with SMEs and interviewees, many highlighted the need for hard and soft skills, along with a blend of experience or a demonstrated ability to learn and perform with high aptitude. Hard skills included technical education, and soft skills included strong communication, interpersonal skills, risk tolerance,

[18] Raymond, 2020.

dedication to mission, and excellence in leadership. The selection process might also implement data-driven processes that identify and select officers and enlisted recruits with the best capabilities to lead the force. Interviewees also advocated for the same selection criteria as employed by NRO and SpRCO.

For civilian personnel, USSF should engage its direct hiring authorities for space-specific career fields and advocate for additional recruitment incentives if they are needed but lack appropriate authorities. USSF could also consider nontraditional industry pathways and rotation programs, direct commissions, and best practices to attract and recruit top technical talent.

New Career Fields and Career Pathways

Embracing a "we are all space warrior engineers" identity and culture means reimagining previously siloed career fields that are the legacy of U.S. Air Force space operations. For space acquisition to truly be a warfighting capability, USSF personnel should understand both the space acquisition process (in any form) and the nature of space operations.

For military personnel, this effort includes initiatives to create modern career fields unique to USSF that attract the desired combination of expertise, mindset, and passion. These would be hybrid career fields that reorganize and redefine the existing acquisition, operator, cyber, and intelligence communities. This initiative could also include officer and enlisted service members, along with different career pathways that involve a variety of possible specializations and associated opportunities for advancement. Merging career fields not only enables greater flexibility in talent management, but it also reinforces a unified culture and mission. Moreover, it limits the ability for one career field to dominate USSF culture, as with pilots in the U.S. Air Force.

For civilians, this effort is also an opportunity to reimagine career pathways through a review of position classifications, assignment, and allocation. This could also be an opportunity to rethink functional communities—the way in which current career codes are managed—to promote greater and more unified alignment across the enterprise.

For both military and civilian personnel, industry rotations should be a stepping stone to career advancement. Industry rotations provide mission-enhancing experience and will enhance USSF through the betterment of its talent in the following ways:

- growing and cultivating expertise in the latest commercial space innovations that could advance USSF's mission
- providing new and challenging opportunities to learn and develop professionally, boosting morale and enthusiasm for the USSF brand and mission
- encouraging more cross-fertilization and collaboration among USSF, industry partners, and the greater space technology community
- creating and maintaining valuable partnerships with industry that helps them better understand USSF's needs and mission goals.

A critical part of these industry rotations is flexibility in structure and timing. Depending on the venue, nature of partnership, and career path of the space professional and the goals of USSF, a rotation could range from a few months to two years. Moreover, a civilian or officer could have multiple rotations across different industry partners during their career. Therefore, this industry rotation program would be unique to USSF and not an extension of existing U.S. Air Force programs such as Education With Industry[19] or AF Ventures.[20] The rotation would count as time in service/grade, and it would also be unique in that it is strongly encouraged, or even required, for promotion to general officer or Senior Executive Service positions.[21]

Interviewees noted different degrees to which military career fields could be merged, particularly in the acquirer and operator communities. Options range from having a limited single tour, rotation, or assignment in the other field to completely merging the fields with little differentiation until the warrior engineer chooses to specialize later in their career. Any approach has trade-offs, which should be the subject of further study. Although outside the scope of this study, the feasibility of having analogous or blended enlisted hybrid occupational specialties and a new process for how military personnel are assigned, tracked, and evaluated is also an important consideration for any hybrid approach.[22]

Reevaluate Uniformed Role in Space Acquisitions

Discussions with DAF acquisition SMEs and our interviews revealed several challenges with the current officer role in effectively managing a space acquisition program team:

- officer selection and duration of assignment as it affects future promotion and advancement opportunities
- officers' routine rotation across programs to gain experience in different phases, which enhances officer learning but limits team continuity when billets are reorganized or reallocated
- ability to coordinate across personnel in different units and locations
- variation in officers' level of skills and experience in acquisitions and as a space operator.

Multiple discussions with SMEs and interviewees raised the question of military personnel's role in acquisitions, given that civilians do much of the underlying work and pro-

19 Air Force Institute of Technology, "Education with Industry (EWI) Program), webpage, undated.

20 AF Ventures, homepage, undated.

21 This would be similar to how having a joint assignment is needed for military officers to be promoted to general officer. USSF would need to evaluate whether industry and joint assignments are both required for selection to general officer and Senior Executive Service.

22 For example, one interviewee believed potential changes to occupation specialty codes could provide an opportunity to redefine performance evaluation criteria—for example, by focusing on experiences gained instead of traditional metrics, such as a meeting a project milestone on budget.

vide continuity from end to end in the process. Along with the challenges noted here, the interviewees confirmed that rotating military officers offers the opportunity to bring in new thinking and ensure a refresh of talent. As we rethink how talent is assigned and managed in USSF, it is worth reviewing the future role of acquisition officers in a civilian-led acquisition force. Options include focusing on high-level leadership and oversight (although the downside here would be less hands-on, in-depth involvement and reduced insight into daily activities), "homesteading" to maintain a longer assignment in a way that does not diminish promotion potential, and having military personnel overlap by formalizing on-the-job training to ensure continuity in rotating out. We finally note that having an independent personnel system and new career specialties and pathways could further mitigate some of the identified challenges.

Summary

As a new military service, USSF has a unique opportunity to reinvent talent management as part of reinventing space acquisition as a warfighting capability. This includes creating a new, independent USSF identity and brand and a mission-driven culture that appropriately aligns incentives with desired performance. To truly be "the envy of all services," USSF should transform its approach to talent management by putting talent first: creating new career fields and pathways to promote unity and inclusion; establishing customized rotations with industry to grow and cultivate expertise in the latest innovations; and creating a culture that rewards data-driven thinking, agility, and excellence in communication and leadership.

CHAPTER FIVE

Interactions with Industry and Other Government Agencies

USSF, more than any other military service, depends on technology to achieve its mission and provide space-based capabilities to DoD, IC, other governmental organizations, partner nation governments, and civilians around the world. Although DoD has developed and matured its space-based capabilities, commercial industry has played an increasingly dominant role in the development and proliferation of space-related products and services. In the United States, commercial industry is, or at least is becoming, the new center of gravity for space technology—a role the U.S. government and those of other spacefaring nations have historically held. This shift presents USSF with both new opportunities and challenges, the responses to which will determine its success. In this chapter, we will describe what we learned from the existing literature and interviewees regarding the need for USSF to

- partner effectively with industry, other DoD and IC organizations, civilian organizations, and our allies who have or are developing a presence in space
- focus on enterprise capability rather than individual system development
- properly organize, train, and equip (OT&E) and integrate operational-level acquisition organizations to achieve diverse acquisition outcomes.

Industry Interaction Framing Assumptions

The discussion of these opportunities and challenges in the following sections are predicated on four framing assumptions:

1. A space technical architecture exists to help guide USSF engagement with—and investment in—industry. Without such a road map, USSF may lack the strategic direction and unity of effort to effectively and efficiently engage industry.
2. USSF has the technical expertise to evaluate and leverage industry advancements. Without a deep knowledge base, USSF may not be able to identify emerging opportunities that are viable or current opportunities that are no longer viable.

3. Acquisition mechanisms and funding flexibility enable agile investment and divestment in industry advancements. Without adequate mechanisms and resources, USSF may not be able explore and exploit crucial opportunities.
4. USSF develops a risk-tolerant culture that is able to move assets away from investments that are not going to pay off to ones that are more likely to be successful—all of which will help it to accept the potential failures of rapid innovation and make full use of the expertise, tools, and resources at its disposal. Without greater risk tolerance, USSF may lack a willingness to fully engage with or invest in industry because of the fear of negative consequences. Failure is the only way to push the limits of the possible and advance the state of the art. The ability to rapidly push those limits is critical to USSF's success.

Improving Partnerships with Industry and Other Organizations

U.S. Space Force and Industry

USSF is dependent on commercial industry to develop and produce spaced-based products and services. Thus, USSF will need to ensure that it continues to support the health and growth of the space industrial base[1] and build strong relationships and mutual understanding with traditional and nontraditional companies to partner more effectively. Now more than ever, the national security space enterprise has a symbiotic relationship with industry that requires true partnerships. Thus, USSF should focus on four key areas to improve its engagement with industry and strengthen its relationships.

Broaden Situational Awareness of Industry Developments

Situational awareness of industry and its technological developments is critical to leveraging industry to the maximum extent possible. Multiple USSF organizations (in particular, the Space Labs and SpaceWERX) should have insight into the space industrial base as a whole and at the company level. As several experts noted, to broaden its situational awareness, USSF should increase engagement with industry using both push and pull mechanisms. Push mechanisms include outreach activities in which USSF goes out to industry, such as participating in industry events and organizations (e.g., Catalyst Accelerator, Starburst, and the National Defense Industrial Association[2]) and potentially performing market research and direct outreach to specific companies. Pull mechanisms include engagement activities in which USSF

---

[1] Steven Butow, Thomas Cooley, Eric Felt, and Joel B. Mozer, *State of the Space Industrial Base: A Time for Action to Sustain US Economic & Military Leadership in Space*, Washington, D.C.: Center for Strategic and International Studies, July 2020.

[2] Catalyst and Starburst are accelerators that provide funding and mentorship to startups and small businesses in the space industry. The National Defense Industrial Association is a trade association for the U.S. government and defense industrial base that helps promote better communication and situational awareness and identifies key challenges and opportunities across the base.

brings in potential industry partners, such as through requests for information or proposals to conduct work, pitch days, and challenges. USSF may need dedicated strategies, resources, and personnel to enable a systematic approach to engage with industry and maintain an adequate level of situational awareness. Such an approach to industry engagement should focus on achieving the space architecture and connecting industry partners and solutions with warfighters and their needs and also on exploring opportunities that could lead to significant leaps in technological development and warfighter capabilities and advantages.

Improve Two-Way Communication with Industry

USSF also needs to improve two-way communication with industry to build stronger relationships and develop mutual understanding and trust. USSF should communicate to industry its values and needs both in the near term and long term. Specifically, USSF should communicate its overall strategy and goals for industry, the space technical architecture, requirements (proportional to risk),[3] and other needs (e.g., administrative, reporting, and cost accounting requirements). Mechanisms to enhance USSF communication with industry include increasing the frequency of industry days; expanding participation in industry days; and providing regular written reports to industry on the status of USSF strategies, plans, and needs.

Conversely, industry needs to communicate to USSF the current state of technology and the state of the possible. Mechanisms for industry to more regularly communicate technical advancements to USSF include increasing the frequency of pitch days; expanding participation in pitch days; and developing processes to receive, route, review, and respond to unsolicited white papers regarding new or evolving technologies with DoD applications. Industry could also provide feedback on the space technical architecture and requirements to improve architecture and system designs. Moreover, USSF needs to understand how the various industry segments operate, how they are incentivized, and what they need to remain competitive in the market. USSF would also benefit from understanding the conditions under which commercial companies would be able and willing to partner and better align with USSF goals. However, USSF will likely need to demonstrate commitment to these goals and incentivize companies either in the near term or long term. Incentives could be financial (e.g., R&D funding or potential contracts) or nonfinancial (e.g., reduced reporting or cost accounting requirements or favorable intellectual property or data rights agreements) and likely would include tailored support to nontraditional companies.

We note that, currently, certain acquisition mechanisms restrict continuous and transparent communication with industry stakeholders at certain times. However, a transformed USSF space acquisition process could reexamine these communications and existing safeguards. Regardless, as we heard several times, USSF can and should be less reserved in its

[3] For example, systems early in the R&D cycle or noncritical subsystems or components may not warrant stringent requirements that could lead to unnecessary and significant increases in cost and schedule. Conversely, systems moving into the production phase or systems with critical missions will likely require more-stringent requirements to ensure that technical maturity and full capabilities are delivered.

formal and informal communication with industry—when and how it is allowed within the law and acquisition regulations.

Another invaluable way to improve situational awareness and communication is by expanding exchange programs with industry, as mentioned in Chapter Four.[4] Through immersion, USSF personnel can quickly learn about industry's diverse environments, cultures, and practices and build stronger relationships. Conversely, USSF personnel can also communicate the service's perspective and needs to industry through these more-informal interactions. In addition, USSF should leverage reserve personnel with experience in industry to the maximum extent possible to further improve situational awareness and gain valuable insights.

Effective communication between USSF and industry is also restricted by the classification of space-related plans, programs, requirements, and initiatives. Reexamining the classification of such information and expanding access when and where appropriate would help improve communication and understanding between USSF and industry. However, a lack of classified information systems within industry also impedes classified information-sharing both prior to and during program execution. Initiatives to help expand the number of classified information systems or provide greater access to these systems for industry partners when and where appropriate would further help improve communication.

Agilely Invest and Divest to Explore Potential Opportunities

Broader situational awareness will help USSF identify potential opportunities within industry. However, USSF will also need to quickly and simultaneously perform other functions to agilely invest and divest to explore these opportunities. Understanding that these are not serially performed, the HSA may use the flexibility that consolidated PEs and minimal colors of money provide to rapidly invest and divest in technologies identified by USSF's technologists. The HSA can also direct them to assess maturity levels of new technology for recommendation to acquirers. With this information, new technologies can be inserted into the right programs at the appropriate point in their life cycle. Once this is done, USSF can use flexible funding and acquisition mechanisms (e.g., other transaction authorities [OTAs] and Small Business Innovation Research) to invest agilely.

Second, USSF will also need processes to quickly develop requirements that are proportional to risk; the stage of development; and the criticality of the system, subsystem, or component. More risk-proportional requirements would enable industry to develop the best possible solution—whether a military or dual-use system or commercial service—and potentially speed up development and reduce costs. Solicitations should also be simple, fast, and unclassified to allow wider industry participation and speed up the solicitation process. Reexamining classification would not only enable greater information-sharing with industry but also could increase competition. Moreover, robust systems engineering must be in place

---

[4] Such programs, for example, include the Air Force Institute of Technology's Education with Industry Program and AF Ventures' Fellowship Program.

to ensure that industry solutions support the space architecture and warfighting doctrine and CONOPs.

To that end, digital engineering has already demonstrated the ability to achieve faster and better outcomes for numerous processes related to acquisition and life-cycle management.[5] The benefits of digital engineering could be realized across the full life cycle of a system, including design and development, assembly, test and evaluation, and operations and sustainment. Developing "digital twins" of new and/or current real-world systems allows rapid iteration on system designs or upgrades and analysis of their effects on warfighting capabilities and total life-cycle costs, as well as the ability to parallelize currently serial processes (e.g., validation and verification) and make normally discrete activities continuous (e.g., authorization to operate). Digital engineering requires authoritative data, models, and infrastructure that, once developed, can be shared easily with industry to propagate and take greater advantage of these tools. Moreover, combining digital engineering with agile software development and open hardware architectures could provide even more benefit to the space acquisition community and space industrial base.

Third, rapid contracting mechanisms are also important to quickly get companies on contract once opportunities have been identified and requirements have been developed. Rapid contracting is important not only for USSF to agilely invest but also for the financial health and commercial viability of startups and small businesses for whom the timing of funding is as important as—and, in some cases, more important than—the amount of funding.

Finally, USSF must be able to continually reevaluate the potential of its investments and then reinvest, hand off partnerships to other organizations, or divest and redirect funds to new opportunities. This could be achieved using single or multiple contracts. Under a single contract, prenegotiated options to on-ramp and off-ramp combined with ongoing contract performance evaluation processes, such as Earned Value Management,[6] could provide the appropriate opportunity and means. Acquisition mechanisms, such as Small Business Innovation Research (SBIR) and OTA, also use multiple contracts in phases to achieve this goal.[7] However, USSF must ensure that the needed flexibility is balanced with the financial and nonfinancial needs of industry to avoid unintended consequences. Moreover, such flexibility may come at an increased cost to USSF that may need to be accepted.

Ensuring Technology Transition

USSF must be able to transition technology from both traditional and nontraditional companies into new or improved operational capabilities. However, technology transition con-

[5] William Roper, "There is No Spoon: The New Acquisition Reality," virtual address at Air Force Association 2020 Virtual Air, Space, and Cyber Conference, Arlington, Va., October 7, 2020.

[6] Office of the Under Secretary of Defense for Acquisition and Sustainment (Acquisition, Analytics and Policy), *Department of Defense Earned Value Management Implementation Guide (EVMIG)*, Washington, D.C.: U.S. Department of Defense, January 18, 2019.

[7] SBIR has Phase I–III contracts, and OTA includes three types: research, prototype, and production.

tinues to be a challenge. Nontraditional companies, in particular, may have fewer transition mechanisms or greater barriers for certain mechanisms. For example, small companies may produce only subsystems or components and require partnerships with prime contractors to incorporate their technology into new space systems. USSF may need to incentivize prime contractors to use new technologies from small companies. Similarly, startup companies may not have the capabilities to manufacture products at scale. After venture capital funding, startups often face the choice of an acquisition/merger or initial public offering to expand operations and generate returns for investors. USSF may also need to ensure that nontraditional companies such as startups can continue to operate in the manner most advantageous to the space industrial base and USSF.

Depending on the status of the technology, USSF should have at least a notional concept for how the technology could be transitioned.[8] This includes linking new or improved technologies with operational needs to secure a committed sponsor, advocate for resources, and encourage widespread adoption. Additionally, USSF should be helping to make connections between industry stakeholders and potentially incentivizing prime contractors to use new or different subcontractors. Many acquisition mechanisms have processes to transition new technologies from R&D into production. For example, OTAs can transition from research and prototype OTAs to production OTAs without having to recompete the contracts; similarly, SBIR grants can transition from Phase I and II for concept development and R&D into Phase III for commercialization and production (although Phase III is funded outside the traditional SBIR program).[9,10] Some barriers to transition may lie outside the domain of acquisition and arise because of the lack of a senior leader "champion," other committed sponsor, validated warfighter requirements, or allocated funding. USSF must ensure that requirements, funding, acquisition, and other needs are met to ensure transition of critical technologies into new warfighting capabilities.

Finally, USSF should also track investments in industry and collect data on the outcomes of these investments to improve outcomes.[11]

---

[8] Note that USSF can separately invest in efforts at Technical Readiness Level 1, without a particular project or transition strategy in mind, to ensure that there is a flow of new science and technology ready for further investment.

[9] Defense Acquisition University, "Prototype OTs," webpage, undated-b.

[10] Defense Acquisition University, "Small Business Innovation Research (SBIR) and Small Business Technology Transfer (STTR)," webpage, undated-c.

[11] GAO, "Small Business Innovation Research: DOD's Program Has Developed Some Technologies That Support Military Users, but Lack Comprehensive Data on Transition Outcomes," statement of Marie A. Mak, acting director of Acquisition and Sourcing Management, before the House Committee on Small Business, House of Representatives, GAO-14-748T, July 23, 2014b.

U.S. Space Force and Other Organizations

USSF will also need to partner with other DoD and IC organizations to achieve their missions and provide space-based services to a wide spectrum of national security, other governmental, civilian, and international users. USSF will need to understand the warfighting requirements of the other military services and IC, in particular the NRO, for which USSF provides space-based and launch services. USSF should also leverage the capabilities of these organizations to help achieve its mission. For example, USSF will need to work closely with the IC to inform the IC of its plans to get the targeted assessments that will ensure that requirements are focused on the threat and enable threat-informed acquisition as well as to develop doctrine and tactics, techniques, and procedures necessary to operate in the contested space domain. Reserve, and to some extent Guard, personnel could also provide USSF with unique knowledge and skills gathered from their professional activities outside DoD, including experience in industry. USSF should leverage their experience for multiple functions across the service, including in R&D, requirements development, acquisition, and operations and sustainment.

USSF should also leverage close relationships with spacefaring allies to use their space-based capabilities or other opportunities, such as combined ventures or hosted payloads, when appropriate. For example, the Combined Space Operations Center was established in 2018 at Vandenberg Air Force Base and hosts allied nations, including the United Kingdom, Canada, and Australia.[12] USSF is also planning to host two payloads of the Enhanced Polar System Recapitalization on two Norwegian satellites to provide secure satellite communications to the Arctic.[13] Expanding initiatives like these should be a key focus area for USSF.

Focusing on the Development of Enterprise Capabilities

USSF's ability to partner effectively with industry and other entities is necessary but not sufficient to develop the enterprise capabilities needed to meet warfighter requirements and achieve USSF's mission. As we discussed earlier in the report, historically, DoD has focused on the development of individual space systems, which has led to the perceived need for exquisite, global solutions and programs of record that frequently run over budget and over schedule. Instead, USSF leadership should manage from an enterprise perspective and focus on capability development. This paradigm shift represents a change in USSF's approach to program coordination and risk management. However, if implemented properly, this change could increase speed, agility, and integration across USSF and allow space-based technologies and services from across the space industrial base to be harnessed as new or improved warfighting capabilities.

---

[12] Joint Force Space Component Command Public Affairs, "Combined Space Operations Center Established at Vandenberg AFB," Vandenberg Air Force Base, Calif., July 19, 2018.

[13] Caleb Henry, "Northrop Grumman to Build Two Triple-Payload Satellites for Space Norway, SpaceX to Launch," *SpaceNews*, July 3, 2019.

A space technical architecture could help codify the enterprise capabilities needed to support the joint forces, which are agnostic to a specific system, technology, or source (e.g., USSF owned and operated system, allied nation system, or commercial service). The space technical architecture needs to be based on space warfighting doctrine and CONOPs, as well as being adaptive to evolving threats and changes in technology. (This is not to imply this work is not being done. The connection to space doctrine, as it is being developed, and CONOPs is essential in the development of the technical space architecture.) The space architecture could also help guide USSF engagement and investment in industry, as well as provide industry with a road map of USSF's plan and allow industry to better align with the plan if it sees worthwhile opportunities or is otherwise incentivized to do so. The space technical architecture would need to have sufficient scope and detail and be disseminated widely enough to both internal USSF and external stakeholders to have the desired effect.

Focusing on the enterprise rather than specific systems, technologies, or sources can drive competition and expand the space industrial base to improve acquisition outcomes and cost, schedule, and performance. This strategy is distinct from the current strategy, which focuses on long-term relationships with a few large contractors. USSF will need to carefully weigh how these strategies benefit or detract from the enterprise and make improving it the sole priority.

Finally, capability development places a greater emphasis on the synchronized development, production, and fielding of systems to provide capabilities to warfighters. USSF will need to better incentivize companies to deliver synchronized products and services, especially when different capability components are developed by different companies and, thus multistakeholder collaboration and coordination are required.

Achieving Diverse Acquisition Outcomes

When standup is complete, USSF will have an initial set of diverse operational-level organizations that are and will continue to be important for building a more-capable and more-resilient space architecture using the varied strengths of the total workforce.[14] These operational-level organizations include

- **SMC:** traditional development and acquisition of space-related military systems and launch services
- **SpRCO:** rapid development and fielding of critical space capabilities[15]
- **SDA:** development of a proliferated LEO, multiple capability space architecture[16]

[14] USSF structure has been highlighted in numerous articles (see, for example, Lynn Kirby, "USSF Field Command Structure Focuses on Space Warfighter Needs," *Dayton Daily News*, July 2, 2020.

[15] SpRCO, "Space Rapid Capabilities Office," fact sheet, October 28, 2020.

[16] Space Development Agency, "About Us," webpage, undated.

- **Space Labs:** traditional R&D, including partnerships with industry
- **Commercial Satellite Communication Office (CSCO):** acquisition of commercial SATCOM services[17]
- **30th and 45th Space Wings:** launch enterprise.

These organizations provide important developmental and/or acquisition capabilities to USSF. Thus, USSF and SSC should continue to organize, train, and equip the organizations within the service to execute their specialized activities, recognizing that the structure, processes, and culture of these organization may be very different based on their activities.

However, SSC should strive to integrate acquisition activities across these operational-level organizations. USSF could compete acquisition of different capabilities (where appropriate) between these organizations to achieve better acquisition and operational outcomes. For example, competing the acquisition of a new broadband SATCOM capability via a small number of larger satellites in geosynchronous orbit and a large number of smaller satellites in LEO between SMC legacy organizations and SDA is an option. Similarly, USSF could also internally compete the acquisition of SATCOM via military systems and commercial services through SMC legacy organizations and CSCO. This type of competition could increase innovative acquisition approaches and forestall any complacency driven by the thinking of "this is how we've always done it."

USSF has needed and will continue to need organizations dedicated to the development and acquisition of the next evolution of space systems and revolutionary new space systems.[18] Organizations dedicated to more-radical development tend to be smaller, independent organizations with a strong mission focus. Such organizations also tend to pivot frequently as opportunities rise and fall and, by the nature of their mission, require long timelines to achieve their ultimate goals.[19] These characteristics are common within innovation ecosystems but often make these organizations the target of budget cuts when the operational needs of today outweigh the strategic needs of tomorrow. Thus, USSF must be committed to protecting and advocating for these organizations and providing them with the time, resources, and flexibility needed to achieve their goals.

At the strategic level, SSC can also play a pivotal role in defining a strategy and goals for USSF partnerships with industry, as well as policies for how operational-level organizations can engage, communicate, and partner with industry more effectively. SSC could engage with the space industrial base and industry executives to ensure (1) the health and growth of the

[17] Courtney Albon, "Space Force to Complete COMSATOM Acquisition Strategy This Summer," *Inside Defense*, March 12, 2020.

[18] GAO, *Defense Science and Technology: Adopting Best Practices Can Improve Innovation Investments and Management*, GAO-17-499, June 2017b.

[19] Shirley M. Ross, Sandra Kay Evans, Lisa Pelled Colabella, and Samantha E. DiNicola, unpublished RAND Corporation research, 2018.

industrial base and (2) USSF's ability to achieve the space architecture and its strategy and goals through its partnerships with industry.

Summary

The standup of USSF and the shift in the technological center of gravity to the industrial base present USSF with new opportunities and challenges, and it must respond accordingly. First, USSF will need to partner more effectively with industry by broadening its situational awareness of technological developments, improving two-way communication, agilely investing and divesting to explore opportunities, and better ensuring technology transition. Second, USSF will need to focus on capability development at the enterprise level and should be guided by a space technical architecture and emerging space doctrine and CONOPs. Third, USSF will need to properly conduct OT&E and integrate operational-level acquisition organizations—guided by new strategies, policies, and goals for industry—to achieve diverse acquisition outcomes. As USSF continues to move forward, it will also need to fully internalize its increasingly symbiotic relationship with the industrial base and prioritize decision-making that carefully balances its needs with the long-term health and growth of the industrial base to fully realize its goals.

CHAPTER SIX

Implementing the Clean Sheet

As a new service, USSF has the opportunity to develop and manage a new culture and new processes appropriate to its small size and technology reliance. It benefits from the lessons learned across and throughout the history of the broader DAF, both the good and the bad, which it can bring to bear on a new approach. It has the opportunity to pick and choose the best-of-breed acquisition features to accomplish its mission. Rather than proposing minor fixes to the traditional U.S. Air Force acquisition structure, we propose a version of USSF acquisition where a single community of warrior engineers views its roles as shaping, managing, and using necessary technology to achieve desired effects. Budget independence from the U.S. Air Force and budget flexibility from having fewer PEs and color-of-money constraints will allow dollars to be applied to USSF's highest priorities—and allowing USSF to flexibly disinvest from less useful programs will support this goal. This vision also encompasses working closely with suppliers and using an approach that allows for effective information exchange with digital engineering as an enabler, while observing laws and regulations, taking advantage of their flexibility, and in some cases, if need be, seeking to change them.

This kind of change needs to be managed carefully. In this chapter, we provide lessons from formal change management and insights from the SMC 2.0 transition, a previous reorganization effort designed to help accelerate space acquisition. SMC 2.0, conceived and initially implemented in 2018 to 2019, involved the same organizations that would have to implement the changes proposed in previous chapters. Our interviews found that SMC's recent experience with those methods provides useful insights into the challenges that USSF would face in the application of similar methods in the future.

Instantiating the Change

The previous chapters build the case for why USSF must depart significantly, in a coordinated enterprisewide fashion, from how the U.S. Air Force has acquired systems and services relevant to space operations in the past. In the early 1990s, large, complex, technologically sophisticated organizations began to realize that change of this magnitude is more likely to yield the desired outcomes if these organizations use a set of formal change management practices to implement the changes.[1] Looking forward, USSF is likely to benefit from using similar practices to ensure implementation of a qualitatively new approach to acquiring systems and services.

In this section, we briefly summarize what formal change management is and the DAF's recent experience with formal change management during the implementation of SMC 2.0, based on discussions with SMEs and stakeholders. We also draw some lessons learned from that experience about how USSF should approach formal change management in the future.[2]

Defining Formal Change Management

Formal change management proceeds in four broad steps:

- Plan for the change.
- Execute the change plan.
- Evaluate the progress and impact; update approach as required.
- Sustain the change and ensure that it is anchored over the long term as standard practice in the organization.

These steps and many of the details associated with them are closely analogous to the steps the DAF has routinely used to acquire major new weapon systems. Broadly speaking, the plan includes the identification and documentation of a gap in capability to be filled (i.e., a requirement). It brings together a coalition to approve this requirement and then resource it. It builds an acquisition strategy that determines the best way to choose a provider of this new system—such as a constellation of satellites—which then designs a detailed plan for how to develop, produce, and sustain the satellites. Once this plan is approved, "execution" first develops the new system and detailed plans on how to produce and sustain it. Execution then steps through the development and production of the satellites. Meanwhile, an arrangement to sustain the new constellation is developed and implemented—for example, a set of ground stations and a command and control structure. When newly produced satellites are available

[1] These practices have been described by many advocates of formal change management. The best known and perhaps most broadly emulated advocate remains John P. Kotter. (For a succinct description of his approach, see John P. Kotter, *Leading Change*, Boston: Harvard Business School Press, 1996.) Different advocates emphasize different practices, but when viewed together, these advocates recommend a remarkably consistent set of practices. (For a concise summary of this consensus, see Frank Camm, Laura Werber, Julie Kim, Elizabeth Wilke, and Rena Rudavsky, *Charting the Course for a New Air Force Inspection System*, Santa Monica, Calif.: RAND Corporation, TR-1291-AF, 2013.) A critical part of this consensus is that organizations seeking to change are likely to fail if they pick and choose among the recommended practices. Each practice builds on the others to create holistic effects that usually lead to success only when an organization accepts the practices as an integrated package.

[2] Our discussion of SMC 2.0 draws heavily on our interviews, on U.S. Air Force and McKinsey & Company documents not available to the public, and on a series of documents RAND produced while offering technical support to the SMC 2.0 effort.

to operate, they fall in the aforementioned new sustainment arrangement, which ensures that they can operate in a way that adds to the new mission capability. Although this approach is typically designed to be linear, surprises inevitably arise as the DAF drives through the life cycle, requiring adjustments in the overall plan to overcome new challenges during development and production or to ensure that a new constellation of satellites can address whatever the threat turns out to be when the constellation finally becomes operational.

Formal change management uses analogous steps to produce some new version of the organization rather than create a new hardware or software system. Where a weapon system acquisition gives a great deal of attention to the engineering performance of the new system and its reliability and availability during operations, formal change management emphasizes how organizational processes and the behavior of the organization's personnel must change to ensure that the change achieves the desired improvement in organizational performance.

Formal Change Management in SMC 2.0

SMC 2.0 was rooted in U.S. Air Force leadership's belief that there was a need to accelerate space acquisition. Because SMC was the main acquisition arm for space equipment, a decision was made to transform SMC into a more agile organization. The U.S. Air Force was introduced to important elements in the first two steps of formal change management described earlier when it contracted with McKinsey & Company to help guide the implementation of SMC 2.0 in 2018 to 2019. At the beginning of this relationship, the general perception within SMC was that it knew best how to acquire space systems and services, while McKinsey knew best how to change organizations. Each side brought its own core competency to SMC 2.0 over the course of the initiative's implementation.

Discussions revealed that SMC approached its transition to SMC 2.0 with a clear vision of the desired change and the motivation for that change.[3] McKinsey began its engagement by assessing the as-is "organizational health" of SMC—the characteristics of its processes and personnel and their likely ability to sustain the change that SMC envisioned.[4] The team then assembled a coalition of relevant stakeholders and set up a governance structure that these stakeholders could use to work together to pursue change. Specific change targets were then refined to first reach IOC (achieved in October 2018) and then to full operating capability

[3] Broadly speaking, SMC 2.0 sought to balance a set of new goals that included the following: (1) use a more formal vision of the enterprisewide priorities of the organization to lead SMC, (2) reduce stovepipes that had formed over time around individual programs, (3) give greater emphasis to the phases of acquisition, within which specialists could reallocate their resources across programs as the programs passed through them over time, and (4) delegate much greater decisionmaking authority to the leaders responsible for their respective phases of acquisition. For details, see SMC, *Implementation Plan for SMC Space Acquisition Re-Architecture*, Los Angeles Air Force Base, Calif.: U.S. Air Force, March 2018.

[4] McKinsey reported its measurements of the organizational health of SMC to SMC in 2018. For a general description of the organizational health index that McKinsey used, see Chris Gagnon, Elizabeth John, and Rob Theunissen, "Organizational Health: A Fast Track to Performance Improvement," *McKinsey Quarterly*, September 2017.

(FOC) in November 2019. Key factors to establishing the change management plan included metrics to measure progress and an improved communication plan to ensure that all participants in SMC 2.0 understood (1) what change was occurring elsewhere and (2) how their pieces of SMC 2.0 fit into the larger effort. During execution of the plans implemented in early 2018, SMC reassigned several key personnel from their everyday mission responsibilities to instantiate SMC 2.0. McKinsey helped SMC train participants on their roles during the implementation and then map a series of incremental successes that SMC 2.0 could build on to demonstrate progress and sustain support for continuing change. SMC and McKinsey collaborated closely over the course of McKinsey's engagement as they learned from each other and mutually adjusted the way forward.

Interviews revealed some senior staff concerns when SMC 2.0 FOC was declared a month early, because the stated goals for FOC had not been achieved. However, top leadership was focused on change being "a journey, not a destination"[5] and was concerned that focusing too much on details would distract SMC from a longer-term agenda of change. Part of this vision was an agile approach to prepare SMC for a dynamic environment in which neither the refinement of weapon systems nor change in an organization is ever complete. This was different from the traditional "waterfall" vision in which requirements "fall down" into development, then production, and then sustainment with little or no ability to adjust the plan. In an agile setting, formal change management does not end; rather, it must be integrated as an essential element of USSF's ongoing mission. Under this approach, SMC 2.0 was just the kickoff.

Elements of Formal Change Management to Focus on the Future

Ensuring that the holistic changes are permanent requires formal change management to pursue the wholesale changes recommended in this report. The experience of SMC 2.0 suggests several elements of formal change management to which USSF should give special attention.

Leadership

The sustained support of the relevant leadership is critical to the success of any major organizational change effort. Sustaining this support can be challenging in a defense setting where the military leadership team is constantly turning over. Fortunately, DoD has learned how to sustain support for weapon system acquisitions that continue through multiple leadership teams. To ensure the success of future change efforts, USSF must invest as much attention to organizational change as it does to major new weapon system acquisitions that continue through multiple leadership teams. For example, SMC leadership played an active role in SMC 2.0 to support its success, and future leaders will need to recognize this to ensure that successful change continues.

[5] SMC Public Affairs, "E.P.I.C. Speed Ahead—SMC at 'Full Operating Capability,'" November 22, 2019.

Our interviewees identified strong interest for such a senior commitment to continue. Interviewees also identified concern among those committed to continuing change that (1) future leadership teams will not sustain the commitment that led to SMC 2.0 and (2) the time costs imposed on participants in SMC 2.0 during its implementation may discourage them from supporting a comparable level of commitment over the longer term. We say more about this next.

Mission Clarity

Two issues associated with mission clarity are important for USSF.

First, a commitment to make organizational change or agility an integral part of USSF's mission will require a formal explanation of this commitment. For example, how much effort should USSF personnel invest in pursuing organizational change relative to the time they devote to their day-to-day responsibilities? What kinds of activities are appropriate? Will their performance in continuing change efforts receive as much attention in their performance reviews as the performance of their day-to-day responsibilities does? USSF will need to clarify how much commitment to change is enough and then communicate these priorities clearly.[6]

Second, the previous chapters call for fundamental changes in organizational culture. Culture is a notoriously hard thing to define in practical terms and even harder to measure. But for the purposes of discussion, consider four different elements of culture that are relevant to the discussion in the previous chapters. In the future, USSF might choose to encourage its acquisition personnel to

- become warrior engineers who treat space as militarily contestable
- recognize that delay could impose mission risks that are worse than traditionally highlighted acquisition risks, such as the potential for cost growth
- share available data and apply shared data to support decisions
- be accountable for results they achieved rather than for their effort.

Without making any judgments about which, if any, of these elements to favor, we simply observe that USSF leaders must be clear about the following:

- what potential culture the leaders seek to instantiate
- how they wish to define and measure each of these elements of culture
- how they wish to weight these elements of culture relative to one another
- how they want to reflect measures of such elements of culture in the operational performance reviews of their personnel.

[6] As a point of reference, 3M gives a high priority to product innovation. It measures what percentage of its net income flows from products introduced in the past three years and judges its employees on the basis of what it observes. It also provides its employees with free time each week to pursue their own initiatives in the hope that this time away from their day-to-day responsibilities might generate ideas for new products.

When trying to change culture, USSF needs to be as clear as possible about what it wants so that its personnel understand the goals and expectations.

Motivation

Personnel in large, complex organizations typically face two different sources of motivation.[7] *Intrinsic motivation* flows from their professional values and their inherent commitment to their organization's mission, both typically buoyed by effective leadership. *Extrinsic motivation* comes from tangible incentives, largely embedded in the performance review process—for example, the effects of awards and other formal recognition, promotions, access to desirable assignments and training, and monetary compensation.

Interviewees noted that, during the transition to SMC 2.0, SMC's leadership promoted the need for change and articulated the new professional values that would be important to sustaining the change. However, the SMC 2.0 change process has not continued long enough to anchor these new values for most of the personnel in the SMC workforce. Also, the performance review process has not yet been adjusted to reflect the new values. Looking forward, USSF will need to ensure that both sources of motivation recognize the goals of organizational change, reward personnel for promoting effective change, and sanction personnel for resisting change. For example, it is one thing to extol the value of taking more risks in decisionmaking, but it is another thing entirely to demonstrate that risk taking[8] (especially if done smartly but ultimately unsuccessfully) will be rewarded. The same applies for any new behavior from an organizational change.

Enablement

Effective motivation will promote change only if personnel have the freedom, skills, and resources to pursue change effectively. SMC 2.0 revealed that one of the most important resources relevant to change was the time a person would spend to participate effectively in change activities. In recognition of this, during SMC 2.0, SMC relieved key participants of some of their normal duties. But the number of these key participants was small. For example, working groups rarely had more than a few members. And senior personnel maintained their responsibilities, although they participated in regular, often weekly, change management

---

[7] For concrete discussions of what the elements of motivation, enablement, and metrics and performance measurements discussed here look like in six different public policy settings, see Brian M. Stecher, Frank Camm, Cheryl L. Damberg, Laura S. Hamilton, Kathleen J. Mullen, Christopher Nelson, Paul Sorensen, Martin Wachs, Allison Yoh, Gail L. Zellman, and Kristin J. Leuschner, *Toward a Culture of Consequences: Performance-Based Accountability Systems for Public Services*, Santa Monica, Calif.: RAND Corporation, MG-1019, 2010.

[8] There need to be clear standards for this risk-taking to accept early failure but drive for more frequent success.

meetings through much of the implementation. Our interviews revealed broad relief among staff that SMC 2.0 was "done" so that they could all return to their primary responsibilities.[9]

If USSF intends to make formal organizational change a part of its personnel's long-term responsibilities, it will need to rebalance the responsibilities of these personnel to ensure that key personnel are enabled to give continuing change appropriate focus and effort.

Metrics and Performance Measurement

Metrics play two roles during an organizational change effort: benchmarking and incentivization.

First, metrics measure the "state of the world" at any point in time so participants understand how they are performing relative to goals and, in particular, whether the initial plan for change is yielding the benefits anticipated. Such tracking helps change managers adjust their plans as data accumulate about the change effort. It also gives change leaders evidence that they can use to build support for further change and to decide which new change activities deserve the greatest emphasis. That is, metrics give all participants actionable language they can use to communicate with one another and support decisions. Precise metrics are especially useful during ongoing organizational change, because participants cannot rely on their historical experience to judge what actions are appropriate. Precise metrics improve their ability to work together to describe the desired future and to coordinate their efforts to achieve that future.

Second, metrics provide the basis for an organization seeking change to apply extrinsic incentives that motivate personnel to pursue as precisely as possible the change the organization seeks. This use of metrics is especially important in an organization seeking to advance enterprisewide goals by delegating authority and responsibility down into the organization so it can move rapidly toward its enterprisewide goals without extensive coordination through bureaucratic hierarchies.

These two roles of metrics are in tension with one another. Incentivizing metrics typically create perverse incentives for personnel to misreport the metrics' values to enhance their own rewards. Unfortunately, misreported metrics lack the validity that personnel need to coordinate their actions, especially when they are engaged in ongoing organizational change.

During the transition to SMC 2.0, SMC gave high priority to developing metrics. It faced serious challenges from the beginning that it had not overcome by the time FOC was declared. These challenges included the following:

- Personnel had difficulty defining key concepts clearly enough to create metrics that measured them. For example, our interviews indicated that, despite considerable effort, SMC could not define risk well enough to design metrics that might promote appropriate risk-taking.

[9] We note here that the message of "done" received by the workforce is quite different than the message of "change is a journey" intended by leadership.

- Where it could define metrics, SMC often lacked existing sources of the data it needed to populate the metrics it sought to create.
- Where appropriate data existed, those who controlled access to the data were often reluctant to share data that others might use to oversee their efforts.
- Without past experience using metrics to systematically inform decisions, many SMC personnel were not motivated to create a data-based infrastructure that would give data a greater role in decisionmaking.

Looking forward, to make effective use of metrics, USSF will need to face these challenges again and overcome them.

Communication

Change management efforts in large, complex organizations routinely discover that participants are not communicating well enough to sustain successful coordination or, even in the face of success, support for continuing change. Effective communication goes from top to bottom to convey the intentions of the leadership; from bottom to top to ensure accountability, to give leaders the information they need to plan next steps, and to give participants a sense of influence in the ongoing effort; and from side to side to allow horizontal coordination without the time and resource burdens of passing information through vertical hierarchies. Effective communication uses multiple media to ensure that all participants can easily access the information they need to motivate them and for them to act.

SMC discovered the vital role of communication early in the SMC 2.0 transition process when players in various parts of the effort were repeatedly surprised by activities elsewhere. SMC responded quickly to such challenges to enhance communication. Its efforts to enhance communication continued throughout the transition. Looking forward, personnel in USSF can learn from the experience of participants in SMC 2.0, both to anticipate what challenges to expect and to learn how to address these challenges quickly when they come to light.

Managing Resistance to Change Among Long-Term Staff

Our interviews point to significant support among senior leaders to promote ideas like those described in the earlier chapters. Interviewees also suggested—although we did not try to measure this directly—that the newer and more-junior personnel generally welcome the chance to have more authority to shape decisions and may be more willing to adopt new processes. They identified that skepticism is highest among longer-term mid-level staff. Mid-level staff may have seen previous change efforts not have the desired results and thus may be more cautious about investing their efforts. Without the support of mid-level staff, the leadership-supported initiatives will likely fail over the longer term.[10] The challenge of get-

[10] Many researchers see middle managers as the key to successful change. See, for example, Behnam Tabrizi, "The Key to Change Is Middle Management," *Harvard Business Review*, October 27, 2014.

ting these personnel to adopt new processes is highlighted by a term that several interviewees used to describe them—the "frozen middle."

Future change management efforts in USSF should give special attention to these personnel. They need the opportunity to shape leadership goals and expectations and clear opportunities to explain how their talents and teams can be best used to implement the changes needed to achieve those goals. They then need to be supported in their efforts to implement the leadership-supported changes, including the right to redirect USSF efforts as needed to achieve the goals. This, of course, requires that upper management have the metrics to measure progress against goals (not against plans), and it requires that changes to compensation and promotion incentives are implemented to reward not just the progress toward goals but the willingness to pursue those goals. Statements from the leadership regarding the consequences of resisting or accepting changes are unlikely to achieve the culture needed for effective transformation without metrics and incentives.

Summary

All of these elements of formal change management are interdependent. Participants cannot understand what to do without clear leadership guidance and training. Leaders cannot convey their guidance or track evidence that change is worthwhile without effective communication. Effective communication is not possible without metrics. None of these efforts will promote effective change unless participants are motivated to execute the leaders' plans and have been enabled to seek change. As active participants in change, longer-term staff who may resist change must be enabled and motivated to promote change. They should be given information needed to convey leadership's priorities through the whole enterprise. They also need the same visibility of the enterprise that their leaders have to see change through to successful outcomes. To implement formal change management effectively in the future, USSF personnel will need to understand how to use all of these elements of formal change management in an integrated manner. Experience from SMC 2.0 can give them valuable information on how to do this.

CHAPTER SEVEN

Conclusions

As we have described in this report, the new clean sheet space acquisition approach and culture require increased agility, rapid decisionmaking, and integration with space operations to enable development and delivery of threat-focused, innovative space capabilities to maintain and strengthen the United States' advantage in space. We offered a clean sheet vision that views and manages acquisition as a warfighting capability. USSF needs independence in budgeting, requirements development, talent management, and acquisition execution to use the following features, which support USSF's vision and provide threat-informed capability on an operationally viable schedule within cost constraints:

- dissolving seams traditionally separating operators and acquirers so that all understand both technology and operations; operators will know how technology flows and changes, and acquirers will know how technology is implemented
- creating an adaptive technical architecture, based on warfighting doctrine and CONOPs, to serve as a framework for decisionmaking and road map for innovation
- establishing a single space acquisition decisionmaker flexibly managing the enterprise—focusing resources on the highest priorities with the authority to invest or divest programs to achieve the enterprise's best value and drive capability synchronization while radically delegating to empowered, expert subordinates
- ensuring a workforce of experts cultivated through selective recruiting, assignments, training, and promotions to be risk tolerant, flexible, collaborative, and enterprise-focused—providing capabilities, not merely systems
- building internal and external outreach mechanisms emphasizing information-sharing, metrics, strong relationships, and mutual trust within and across Congress, DoD, USSF, the IC, other federal agencies, and industry
- fostering a trusting, collaborative relationship with industry—for example, providing industry with a technology road map with innovation on-ramps to accept emerging technology or address changing threats and divestiture off-ramps for obsolete capability.

These components build on each other and contribute to the overall vision depicted in Figure 7.1.

This is a systematic, comprehensive, and holistic approach that, taken together, provides for a transformative approach to space acquisition. USSF leadership must resist treating this

FIGURE 7.1
Components of a Clean Sheet Space Acquisition Enterprise

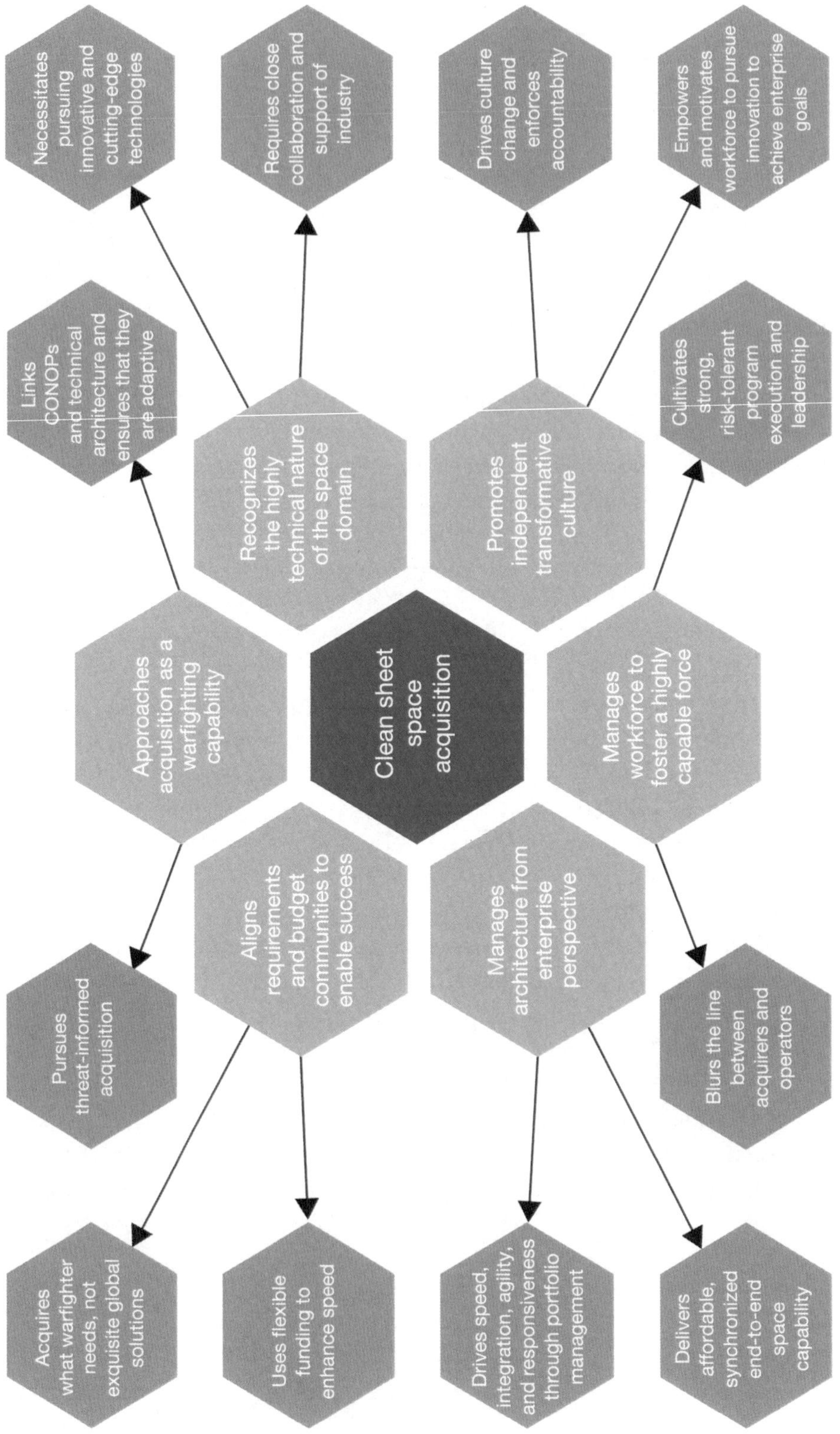

approach as a menu from which to pick and choose. Doing so could achieve small pockets of improvement but not the overall development of culture and effectiveness possible from a holistic implementation. USSF needs the flexibility and authority to invest in *all* of these changes across the enterprise—and Congress will need to ensure that USSF has and maintains the required authorities, including enhanced funding flexibility to allow for investments and disinvestments as the architecture evolves.

The transition to the new acquisition approach described in this report will benefit from and contribute to a new USSF culture that values speed, agility, and capability delivery. These changes must be done with attention to principles of change management and building on the lessons learned from the transition to SMC 2.0. Finally, we close by reemphasizing that these recommendations should be done simultaneously and intentionally to create the right culture and ensure effective and lasting change to establish an acquisition system that is, in the first CSO's vision, threat focused and fueled by innovation—and the envy of all other services.

Abbreviations

AEHF	Advanced Extremely High Frequency
AFCS	Air Force Civilian Service
AFPC	Air Force Personnel Center
AFRS	Air Force Recruiting Service
CDR USSPACECOM	Commander, U.S. Space Command
CONOP	concept of operations
CSCO	Commercial Satellite Communication Office
CSO	Chief of Space Operations
DAF	Department of the Air Force
DAS	Defense Acquisition System
DoD	U.S. Department of Defense
DoN	U.S. Department of the Navy
EMD	Engineering and Manufacturing Development
FFRDC	federally funded research and development center
FOC	full operating capability
GAO	U.S. Government Accountability Office
GPS	global positioning system
HSA	Head of Space Acquisition
IC	Intelligence Community
IOC	initial operational capability
JMS	Joint Space Operations Center Mission System
LEO	low earth orbit
NASA	National Aeronautics and Space Administration
NDAA	National Defense Authorization Act

NRO	National Reconnaissance Office
OCX	Operational Control System
OT&E	organize, train, and equip
OTA	other transaction authority
PE	program element
R&D	research and development
SAC	Space Force Acquisition Council
SATCOM	satellite communications
SBIR	Small Business Innovation Research
SBIRS	Space-Based Infrared System
SDA	Space Development Agency
SMC	Space and Missile Systems Center
SME	subject-matter expert
SpOC	Space Operations Command
SpRCO	Space Rapid Capabilities Office
SSC	Space Systems Command
STARCOM	Space Training and Readiness Command
USSF	U.S. Space Force

References

AF Ventures, homepage, undated. As of April 22, 2021:
https://af-ventures.com/

Air Force Institute of Technology, "Education with Industry (EWI) Program), webpage, undated. As of April 22, 2021:
https://www.afit.edu/CIP/page.cfm?page=1567

Air Force Space Command, "Air Force Space Command History," webpage, undated. As of April 22, 2021:
https://www.afspc.af.mil/About-Us/AFSPC-History/

Air Force Space Command Public Affairs, "Hyten Announces Space Enterprise Vision," April 13, 2016. As of April 22, 2021:
https://www.af.mil/News/Article-Display/Article/719941/
hyten-announces-space-enterprise-vision/

Albon, Courtney, "Space Force to Complete COMSATOM Acquisition Strategy This Summer," *Inside Defense*, March 12, 2020.

Arena, Mark V., Irv Blickstein, Abby Doll, Jeffrey A. Drezner, James G. Kallimani, Jennifer Kavanagh, Daniel F. McCaffrey, Megan McKernan, Charles Nemfakos, Rena Rudavsky, Jerry M. Sollinger, Daniel Tremblay, and Carolyn Wong, *Management Perspectives Pertaining to Root Cause Analyses of Nunn-McCurdy Breaches,* Volume 4: *Program Manager Tenure, Oversight of Acquisition Category II Programs, and Framing Assumptions*, Santa Monica, Calif.: RAND Corporation, MG-1171/4-OSD, 2013. As of April 23, 2021:
https://www.rand.org/pubs/monographs/MG1171z4.html

Baumgartner, Natalie, "Build a Culture That Aligns with People's Values," *Harvard Business Review*, April 8, 2020. As of April 22, 2021:
https://hbr.org/2020/04/build-a-culture-that-aligns-with-peoples-values

Brown, Charles Q., Jr., *Accelerate Change or Lose*, Washington, D.C.: U.S. Air Force, August 31, 2020. As of April 22, 2021:
https://www.af.mil/Portals/1/documents/csaf/CSAF_22/
CSAF_22_Strategic_Approach_Accelerate_Change_or_Lose_31_Aug_2020.pdf

Butow, Steven, Thomas Cooley, Eric Felt, and Joel B. Mozer, *State of the Space Industrial Base: A Time for Action to Sustain US Economic & Military Leadership in Space*, Washington, D.C.: Center for Strategic and International Studies, July 2020. As of April 22, 2021:
http://aerospace.csis.org/wp-content/uploads/2020/07/
State-of-the-Space-Industrial-Base-2020-Report_July-2020_FINAL.pdf

Camm, Frank, Laura Werber, Julie Kim, Elizabeth Wilke, and Rena Rudavsky, *Charting the Course for a New Air Force Inspection System*, Santa Monica, Calif.: RAND Corporation, TR-1291-AF, 2013. As of April 23, 2021:
https://www.rand.org/pubs/technical_reports/TR1291.html

Crutchfield, Dean, "Happy Birthday to America's Most Enduring Brand: The Marines," *Forbes*, November 14, 2011. As of April 22, 2021:
https://www.forbes.com/sites/deancrutchfield/2011/11/14/
happy-birthday-to-americas-most-enduring-brand-the-marines/?sh=6b47737c3df7

Defense Acquisition University, "DAU Adaptive Acquisition Framework," webpage, undated-a. As of April 22, 2021:
https://aaf.dau.edu/aaf/aaf-pathways/

———, "Prototype OTs," webpage, undated-b. As of April 22, 2021:
https://aaf.dau.edu/aaf/contracting-cone/ot/prototype/

———, "Small Business Innovation Research (SBIR) and Small Business Technology Transfer (STTR)," webpage, undated-c. As of April 22, 2021:
https://aaf.dau.edu/aaf/contracting-cone/sbir-sttr/

Defense Intelligence Agency, *Challenges to Security in Space*, Washington, D.C., January 2019. As of April 22, 2021:
https://www.dia.mil/Portals/27/Documents/News/Military%20Power%20Publications/Space_Threat_V14_020119_sm.pdf

Erwin, Sandra, "SMC 2.0: Air Force Begins Major Reorganization of Acquisition Offices," *SpaceNews*, April 17, 2018. As of April 23, 2021:
https://spacenews.com/smc-2-0-air-force-begins-major-reorganization-of-acquisition-offices

Gagnon, Chris, Elizabeth John, and Rob Theunissen, "Organizational Health: A Fast Track to Performance Improvement," *McKinsey Quarterly*, September 2017. As of October 28, 2020:
https://www.mckinsey.com/business-functions/organization/our-insights/organizational-health-a-fast-track-to-performance-improvement

GAO—*See* U.S. Government Accountability Office.

Heller, Jeanne D., ed., *Project AIR FORCE 1999 Annual Report*, Santa Monica, Calif.: RAND Corporation, AR-7042-AF, 2000. As of April 22, 2021:
https://www.rand.org/pubs/annual_reports/AR7042.html

Henry, Caleb, "Northrop Grumman to Build Two Triple-Payload Satellites for Space Norway, SpaceX to Launch," *SpaceNews*, July 3, 2019. As of April 22, 2021:
https://spacenews.com/northrop-grumman-to-build-two-triple-payload-satellites-for-space-norway-spacex-to-launch/

Hitchens, Theresa, "Wilson: DoD Study Finds 'Exquisite' Satellites Still Needed," *Breaking Defense*, April 9, 2019. As of April 21, 2021:
https://breakingdefense.com/2019/04/wilson-dod-study-finds-exquisite-satellites-still-needed/

Holley, Irving Brinton, Jr., "Some Concluding Observations on Military Procurement," in *United States Army in World War II, Buying Aircraft: Matériel Procurement for the Army Air Forces*, Washington, D.C.: Office of the Chief of Military History, Department of the Army, 1964, pp. 569–572. As of April 22, 2021:
https://history.army.mil/html/books/011/11-2/CMH_Pub_11-2.pdf

Husband, Mark, "Information Paper on Framing Assumptions," U.S. Department of Defense, September 13, 2013. As of April 22, 2021:
https://www.acq.osd.mil/aap/assets/docs/2013-09-13-information-paper-framing-assumptions.pdf

Hyten, John E., *Space Mission Force: Developing Space Warfighters for Tomorrow*, Air Force Space Command, white paper, June 29, 2016. As of April 22, 2021:
https://apps.dtic.mil/sti/pdfs/AD1076952.pdf

Independent Business Association, "The Importance of Business Branding," *Medium*, February 11, 2018. As of April 21, 2021:
https://medium.com/@wearetheiba./the-importance-of-business-branding-350e8dd241ec

Joint Force Space Component Command Public Affairs, "Combined Space Operations Center Established at Vandenberg AFB," Vandenberg Air Force Base, Calif., July 19, 2018. As of April 22, 2021:
https://www.afspc.af.mil/News/Article-Display/Article/1579285/combined-space-operations-center-established-at-vandenberg-afb/

Joint Publication 3-0, *Joint Operations*, January 17, 2017, incorporating change 1, Appendix A, *Principles of Joint Operations*, October 22, 2018.

Kim, Yool, Elliot Axelband, Abby Doll, Mel Eisman, Myron Hura, Edward G. Keating, Martin C. Libicki, Bradley Martin, Michael McMahon, Jerry M. Sollinger, Erin York, Mark V. Arena, Irv Blickstein, and William Shelton, *Acquisition of Space Systems, Volume 7: Past Problems and Future Challenges*, Santa Monica, Calif.: RAND Corporation, MG-1171/7-OSD, 2015. As of April 23, 2021:
https://www.rand.org/pubs/monographs/MG1171z7.html

Kirby, Lynn, "USSF Field Command Structure Focuses on Space Warfighter Needs," *Dayton Daily News*, July 2, 2020. As of April 22, 2021:
https://www.daytondailynews.com/news/local/ussf-field-command-structure-focuses-space-warfighter-needs/Y0n69Tfhw5WW5lg4ZRgZ1M/

Kotter, John P., *Leading Change*, Boston: Harvard Business School Press, 1996.

LinkedIn.com, "Already a Space Pro? Add Army Space Ops Officer to Your Resume," GoArmy.com promotion, October 12, 2020.

Losey, Stephen, "Air Force Aims to Modernize Recruiting Amid Growing Challenges," *Air Force Times*, November 2, 2018. As of April 22, 2021:
https://www.airforcetimes.com/news/your-air-force/2018/11/02/air-force-aims-to-modernize-recruiting-amid-growing-challenges/

———, "Crushing Demands of Job Lead Some Air Force Recruits to Falsify Reports," *Air Force Times*, September 1, 2014. As of April 22, 2021:
https://www.airforcetimes.com/education-transition/jobs/2014/09/01/crushing-demands-of-job-lead-some-air-force-recruiters-to-falsify-reports/

Mattis, Jim, *Summary of the 2018 National Defense Strategy of the United States of America: Sharpening the American Military's Competitive Edge*, Washington, D.C.: U.S. Department of Defense, 2018. As of April 22, 2021:
https://dod.defense.gov/Portals/1/Documents/pubs/2018-National-Defense-Strategy-Summary.pdf

McCurdy, Christen, "Space Force Aims for Lean, Fast Force, Gen. John Raymond Says," *Defense News*, September 15, 2020. As of April 21, 2021:
https://www.upi.com/Defense-News/2020/09/15/Space-Force-aims-for-lean-fast-force-Gen-John-Raymond-says/9721600207551/

Milligan, Susan, "Use Your Company's Brand to Find the Best Hires," *HR Magazine*, September 3, 2019. As of April 22, 2021:
https://www.shrm.org/hr-today/news/hr-magazine/fall2019/pages/hr-uses-company-brands-for-best-hires.aspx

MITRE, "Systems Engineering Guide: Systems of Systems," webpage, undated. As of March 21, 2021:
https://www.mitre.org/publications/systems-engineering-guide/enterprise-engineering/systems-of-systems

Office of the Under Secretary of Defense for Acquisition and Sustainment (Acquisition, Analytics and Policy), *Department of Defense Earned Value Management Implementation Guide (EVMIG)*, Washington, D.C.: U.S. Department of Defense, January 18, 2019. As of April 22, 2021:
https://www.acq.osd.mil/evm/assets/docs/DOD%20EVMIG-01-18-2019.pdf

Office of the Under Secretary of Defense (Comptroller), Chief Financial Officer, *Department of Defense Financial Management Regulation (DoD FMR)*, Washington, D.C., DoD 7000.14-R, May 2019. As of March 21, 2021:
https://comptroller.defense.gov/FMR/

Pawlikowski, Ellen, Doug Loverro, and Tom Cristler, "Space: Disruptive Challenges, New Opportunities, and New Strategies," *Strategic Studies Quarterly*, Vol. 6, No. 1, Spring 2012. As of April 22, 2021:
https://www.airuniversity.af.edu/Portals/10/SSQ/documents/Volume-06_Issue-1/Pawlikowski.pdf

Pope, Charles, "Raymond and Space Force Enter New, Ambitious Phase as U.S. Space Command Changes," Air Force Public Affairs, August 24, 2020a. As of April 22, 2021:
https://www.spaceforce.mil/News/Article/2322445/raymond-and-space-force-enter-new-ambitious-phase-as-us-space-command-changes-l/

———, "Driven by 'A Tectonic Shift in Warfare' Raymond Describes Space Force's Achievements and Future," Air Force Public Affairs, September 15, 2020b. As of May 12, 2021:
https://www.af.mil/News/Article-Display/Article/2348524/driven-by-a-tectonic-shift-in-warfare-raymond-describes-space-forces-achievemen/

Public Law 116-92, National Defense Authorization for Fiscal Year 2020, December 20, 2019.

Purdue University, College of Engineering, "System of Systems (SoS)," webpage, undated. As of March 21, 2021:
https://engineering.purdue.edu/Engr/Research/Initiatives/Archive/SoS

Raymond, John, *Chief of Space Operations' Planning Guidance*, Washington, D.C.: U.S. Space Force, November 9, 2020. As of April 22, 2021:
https://media.defense.gov/2020/Nov/09/2002531998/-1/-1/0/CSO%20PLANNING%20GUIDANCE.PDF

Roper, William, "There Is No Spoon: The New Acquisition Reality," virtual address at Air Force Association 2020 Virtual Air, Space, and Cyber Conference, Arlington, Va., October 7, 2020. As of April 22, 2021:
https://software.af.mil/wp-content/uploads/2020/10/There-Is-No-Spoon-Digital-Acquisition-7-Oct-2020-digital-version.pdf

SMC Public Affairs, "E.P.I.C. Speed Ahead—SMC at 'Full Operating Capability,'" November 22, 2019. As of May 29, 2020:
https://www.losangeles.spaceforce.mil/News/Article-Display/Article/2025004/epic-speed-ahead-smc-at-full-operating-capability/

Snow, Shawn, "The Corps Is Finding New Marines Despite Recruiting Challenges," *Marine Times*, November 2018. As of May 15, 2021:
https://www.marinecorpstimes.com/news/your-marine-corps/2018/11/02/the-corps-is-finding-new-marines-despite-recruiting-challenges/

Snyder, Don, Sherrill Lingel, George Nacouzi, Brian Dolan, Jake McKeon, John Speed Meyers, Kurt Klein, and Thomas Hamilton, *Managing Nuclear Modernization Challenges for the U.S. Air Force: A Mission-Centric Approach*, Santa Monica, Calif.: RAND Corporation, RR-3178-AF, 2019. As of December 10, 2020:
https://www.rand.org/pubs/research_reports/RR3178.html

Society for Human Resource Management, "Aligning Workforce Strategies with Business Objectives," September 2015. As of April 22, 2021:
https://www.shrm.org/resourcesandtools/tools-and-samples/toolkits/pages/aligningworkforcestrategies.aspx

Space and Missile Systems Center, *Implementation Plan for SMC Space Acquisition Re-Architecture*, Los Angeles Air Force Base, Calif.: U.S. Air Force, March 2018.

Space Development Agency, "About Us," webpage, undated. As of April 22, 2021:
https://www.sda.mil/home/about-us/

Space Rapid Capabilities Office, "Space Rapid Capabilities Office," fact sheet, October 28, 2020. As of April 22, 2021:
https://www.kirtland.af.mil/Portals/52/SpRCO%20Fact%20Sheet%20Oct%2028%202020.pdf

SpRCO—*See* Space Rapid Capabilities Office.

Stecher, Brian M., Frank Camm, Cheryl L. Damberg, Laura S. Hamilton, Kathleen J. Mullen, Christopher Nelson, Paul Sorensen, Martin Wachs, Allison Yoh, Gail L. Zellman, and Kristin J. Leuschner, *Toward a Culture of Consequences: Performance-Based Accountability Systems for Public Services*, Santa Monica, Calif.: RAND Corporation, MG-1019, 2010. As of April 23, 2021:
https://www.rand.org/pubs/monographs/MG1019.html

Tabrizi, Behnam, "The Key to Change Is Middle Management," *Harvard Business Review*, October 27, 2014.

U.S. Air Force, "Air Force Careers: Explore Careers and Find Your Purpose," webpage, undated-a. As of April 22, 2021:
https://www.airforce.com/careers/browse-careers/space

———, "The Sky Is Not the Limit," webpage, undated-b. As of April 22, 2021:
https://www.airforce.com/spaceforce

U.S. Army Talent Management, homepage, undated. As of April 22, 2021:
https://www.talent.army.mil

U.S. Department of Defense, "Selected Acquisition Report: Global Positioning System III (GPS III)," December 2018.

———, *Defense Space Strategy Summary*, Washington, D.C., June 2020a. As of April 22, 2021:
https://media.defense.gov/2020/Jun/17/2002317391/-1/-1/1/2020_DEFENSE_SPACE_STRATEGY_SUMMARY.PDF

———, "Face of Defense: Recruiting the Next Generation," webpage, October 2020b. As of May 16, 2021:
https://www.defense.gov/Explore/Features/story/Article/2368018/face-of-defense-recruiting-the-next-generation/

U.S. Government Accountability Office, *Defense Acquisitions: Challenges in Aligning Space System Components*, GAO-10-55, October 2009.

———, "Space Acquisitions: DOD Faces Challenges in Fully Realizing Benefits of Satellite Acquisition Improvements," statement of Cristina T. Chaplain, director of Acquisition and Sourcing Management, before the Subcommittee on Strategic Forces, Committee on Armed Services, U.S. Senate, GAO-12-563T, March 21, 2012. As of April 22, 2021:
https://www.gao.gov/assets/gao-12-563t.pdf

———, "Space Acquisitions: Acquisition Management Continues to Improve but Challenges Persist for Current and Future Programs," statement of Cristina T. Chaplain, director of Acquisition and Sourcing Management, before the Subcommittee on Strategic Forces, Committee on Armed Services, U.S. Senate, GAO-14-382T, March 12, 2014a. As of April 22, 2021:
https://www.gao.gov/assets/gao-14-382t.pdf

———, "Small Business Innovation Research: DOD's Program Has Developed Some Technologies That Support Military Users, but Lack Comprehensive Data on Transition Outcomes," statement of Marie A. Mak, acting director of Acquisition and Sourcing Management, before the House Committee on Small Business, House of Representatives, GAO-14-748T, July 23, 2014b. As of April 22, 2021:
https://www.gao.gov/assets/670/664971.pdf

———, "Space Acquisitions: Challenges Facing DOD as It Changes Approaches to Space Acquisitions," statement of Cristina T. Chaplain, director of Acquisition and Sourcing Management, before the Subcommittee on Strategic Forces, Committee on Armed Services, U.S. Senate, GAO-16-471T, March 9, 2016a. As of April 22, 2021:
https://www.gao.gov/assets/gao-16-471t.pdf

———, "DOD Space Acquisitions Management and Oversight: Information Presented to Congressional Committees," GAO-16-592R, July 27, 2016b. As of April 22, 2021:
https://www.gao.gov/assets/gao-16-592r.pdf

———, "Space Acquisitions: DOD Continues to Face Challenges of Delayed Delivery of Critical Space Capabilities and Fragmented Leadership," statement of Cristina T. Chaplain, director of Acquisition and Sourcing Management, before the Subcommittee on Strategic Forces, Committee on Armed Services, U.S. Senate, GAO-17-619T, May 17, 2017a. As of April 22, 2021:
https://www.gao.gov/assets/gao-17-619t.pdf

———, *Defense Science and Technology: Adopting Best Practices Can Improve Innovation Investments and Management*, GAO-17-499, June 2017b. As of April 22, 2021:
https://www.gao.gov/assets/690/685524.pdf

———, *Defense Space Systems: DOD Should Collect and Maintain Data on Its Space Acquisition Workforce*, GAO-19-240, March 2019a. As of April 22, 2021:
https://www.gao.gov/assets/gao-19-240.pdf

———, "Space Acquisitions: DOD Faces Significant Challenges as It Seeks to Address Threats and Accelerate Space Programs," statement of Cristina T. Chaplain, director of Contracting and National Security Acquisitions, before the Subcommittee on Strategic Forces, Committee on Armed Services, House of Representatives, GAO-19-482T, April 3, 2019b. As of April 22, 2021:
https://www.gao.gov/assets/gao-19-482t.pdf

———, *Weapon Systems Annual Assessment: Limited Use of Knowledge-Based Practices Continues to Undercut DOD's Investments*, GAO-19-336SP, May 2019c. As of April 22, 2021:
https://www.gao.gov/assets/700/698933.pdf

——, *Space Command and Control: Comprehensive Planning and Oversight Could Help DOD Acquire Critical Capabilities and Address Challenges*, GAO-20-146, October 2019d. As of April 22, 2021:
https://www.gao.gov/assets/gao-20-146.pdf

USSF—*See* U.S. Space Force.

U.S. Space Force, "About Space Force," webpage, undated. As of April 22, 2021:
https://www.spaceforce.mil/About-Us/About-Space-Force/

Vedda, James A., *Center for Space Policy and Strategy Policy Paper, National Space Council: History and Potential*, El Segundo, Calif.: Aerospace Corporation, November 2016. As of June 14, 2021:
https://aerospace.org/sites/default/files/2018-05/NationalSpaceCouncil.pdf

Vergun, David, "Space Force Leader Discusses Newest Military Service," *DoD News*, October 27, 2020. As of April 22, 2021:
https://www.defense.gov/Explore/News/Article/Article/2396174/space-force-leader-discusses-newest-military-service/

Young, A. Thomas, Edward Anderson, Lyle Bien, Ronald R. Fogleman, Keith Hall, Lester Lyles, and Hans Mark, *Leadership, Management, and Organization for National Security Space: Report to Congress of the Independent Assessment Panel on the Organization and Management of National Security Space*, Alexandria, Va.: Institute for Defense Analyses, 2008. As of April 22, 2021:
https://apps.dtic.mil/sti/pdfs/ADA486551.pdf

Zimmerman, Philomena, and Darren Rhyne, "Lunch and Learn—Digital Engineering," presentation, Defense Acquisition University, May 23, 2018. As of March 21, 2021:
https://www.dau.edu/events/Lunch-and-Learn---Digital-Engineering

ZipRecruiter, "Why Company Culture Is So Important for Attracting Talent," blog post, March 25, 2015. As of April 22, 2021:
https://www.ziprecruiter.com/blog/why-company-culture-is-so-important-for-attracting-talent/